Praise for

Inquiries into the Nature of Slow Money

"We have to find a new form of economy, an economy that knows how to govern its limits, an economy that respects nature and acts at the service of man, a situation where political and humanistic choices govern the economy and not the other way around. We have to discover new economic relationships that move at a more natural pace. This is the potential of slow money."

—Carlo Petrini, founder of Slow Food

"Woody Tasch has one of those fast minds that always seem to ask the right slow questions. He is on to something: a new vision of deploying capital in a way that might offer a true alternative to faster and faster, bigger and bigger, more and more global. I've been saying for years that we need to feed the soil, not the plant—slow money is about feeding the soil of the economy."

—Eliot Coleman, farmer and author of *The New Organic Grower*, *Four-Season Harvest*, and *The Winter Harvest Handbook*

"Woody Tasch has been warning us for many years about the dangers of fast money, above all regarding the long term damage inflicted by short term greed. Now that the mathematical manure of maximum leverage has hit the fan of finite natural resources, he presents a compelling case for the rich compost of 'slow' money, and for investment criteria that can sustain and preserve the planet's wealth for generations to come. Indispensable reading, to be placed on the same shelf as Berry and Schumacher."

—Gregory Whitehead, Treasurer, The Whitehead Foundation

"This book is an essential read for anyone who is concerned about the human condition and our planet. Few have taken the idea of walking your talk this much to its essence: all the way to where money meets the earth, so that we can begin to build a truly healthy economy."

—Mark Finser, Chair of the Board,
RSF Social Finance

"Every once in a while, an idea comes around that you immediately know is not only a good one, but in fact is an absolutely necessary one. Slow money is such an idea. Money is a powerful thing and whatever we collectively put our money into goes a long way toward creating the world that we live in. So far, those choices have led to many things, including a broken world food system, where nobody knows where their food comes from or what it takes to grow it. To become so divorced from something as essential as our food has had many disastrous consequences. I have great hope that sustainable, locally based food systems will help us all in more ways than we imagine. Slow money can play a huge role in doing this and Woody's book is an inspiration to all of us working in sustainable agriculture. I can't wait to live in a world supported by slow money."

—Tom Stearns, President,
High Mowing Organic Seeds

"The idea of the appropriate velocity of money is a radical idea. We're not talking here about money velocity in terms of traditional econometrics. We are talking about the speed of money in cultural terms, in biological terms, in relation to the speed of water and the speed of information. *Inquiries into the Nature of Slow Money* is charting the course to a new financial-services sector, the nurture capital industry, recognizing and remediating the human and ecological costs of money that is too fast."

—David Orr, author of *Down to the Wire: Confronting Climate Collapse* and professor, Oberlin College

"Becoming conscious of the power of your consumer dollars is key to building healthy local economies. So is becoming conscious of the power of your investment dollars. We need to rebuild healthy relationships within companies and between companies and all stakeholders, including other species. This is and must be a mass movement. [*Slow Money*] provides the language, the vocabulary, and the framing—empowering us to connect our values to our investments and our investments to local food systems. *Slow Money* catalyzes, crystallizes, and inspires on the path to sustainable communities and a sustainable economy."

—Judy Wicks, founder of the White Dog Café
and co-founder, Business Alliance for Local
Living Economies (BALLE)

"What Woody has given us is a deep and creative analysis of how fast money is destroying our planetary home and its inhabitants, and an evolving recipe for slowing down and redirecting our investments so as to enhance human development and mitigate human environmental impact. Sometimes books come along at exactly the right time to help us understand where we were headed. It goes along with *Small Is Beautiful* on my 'books that matter' shelf."

—Joan Dye Gussow, author of *This Organic Life*
and professor emerita, Columbia University

"This book marks the beginning of a movement."

—Triple Pundit

"Tasch's book includes but goes beyond science, just like it includes but goes beyond economics. It has the potential to impact food policy and climate policy in profound ways. Connecting soil, food, and money, really connecting them—this may just be a threshold moment in the history of sustainable agriculture."

—Mardi Mellon, director of Food and Environment
Program, Union of Concerned Scientists

"Everyone should read this book."

—Neil Chrisman, former managing director, J.P. Morgan

"*Inquiries into the Nature of Slow Money* portrays a world I want to live in. This book should be core reading for every MBA student."

—Ed Church, Executive Director,
Institute for Environmental Entrepreneurship

"*Slow Money* is a great step toward new funding strategies that can help small and medium-size farms, farm businesses, and their communities become more sustainable."

—Richard Rominger, former Deputy Secretary,
U.S. Department of Agriculture

"The vision for Slow Money is brilliant. This book is visionary and pragmatic. We should all listen."

—Paul Dolan, former president of Fetzer Vineyards
and author of *True to Our Roots*

INQUIRIES INTO THE NATURE OF

slow money

investing as if
food, farms, and fertility
mattered

WOODY TASCH

CHELSEA GREEN PUBLISHING
WHITE RIVER JUNCTION, VERMONT

Project Manager: Emily Foote
Developmental Editor: Joni Praded
Copy Editor: Cannon Labrie
Proofreader: Ellen Brownstein
Designer: Peter Holm, Sterling Hill Productions

Printed in the United States of America
First hardcover printing October, 2008
First paperback printing May, 2010
10 9 8 7 6 5 4   10 11 12

Our Commitment to Green Publishing

Chelsea Green sees publishing as a tool for cultural change and ecological stewardship. We strive to align our book manufacturing practices with our editorial mission and to reduce the impact of our business enterprise on the environment. We print our books and catalogs on chlorine-free recycled paper, using vegetable-based inks whenever possible. This book may cost slightly more because we use recycled paper, and we hope you'll agree that it's worth it. Chelsea Green is a member of the Green Press Initiative (www.greenpressinitiative.org), a nonprofit coalition of publishers, manufacturers, and authors working to protect the world's endangered forests and conserve natural resources. *Inquiries into the Nature of Slow Money* was printed on Natures Book Natural, a 30-percent postconsumer recycled, FSC-certified paper supplied by Thomson Shore.

Library of Congress Cataloging-in-Publication Data
Tasch, Woody, 1951-
Inquiries into the nature of slow money : investing as if food, farms, and fertility mattered / Woody Tasch.
p. cm.
Includes bibliographical references.
ISBN 978-1-60358-006-9 (hardcover) — 978-1-60358-254-4 (pbk.)
1. Sustainable agriculture--United States. 2. Food industry and trade--United States. 3. Social responsibility of business--United States. I. Title.

S441.T37 2008
332.67'22--dc22

2008043663

Chelsea Green Publishing Company
Post Office Box 428
White River Junction, VT 05001
(802) 295-6300
www.chelseagreen.com

The richer a society, the more impossible it becomes to do worthwhile things without immediate pay-off.

—E. F. SCHUMACHER

You want to talk about returns? At 1,000:1 in four months, a tomato seed makes even the highest fliers seem paltry.

—ELIOT COLEMAN

I prefer butter to margarine, because I trust cows more than I trust chemists.

—JOAN GUSSOW

Money is like manure.

—SIR FRANCIS BACON,
THORNTON WILDER,
J. PAUL GETTY

contents

foreword

In the last 300 years, economics has been dominant over humanism and culture. In the last 100 years, capitalism has become the motor for culture and for politics on a universal scale.

Capitalism and the global economy have reconceived of development as a linear process that has no end. They have conceived of nature as a resource that is infinite. Today we stand as witnesses of the environmental disaster this has caused.

We have to find a new form of economy, an economy that knows how to govern its limits, an economy that respects nature and acts at the service of man, a situation where political and humanistic choices govern the economy and not the other way around. We have to discover new economic relationships that respect the pace of nature. Lacking this respect, we have an economy that is speeding out of control, promoting unlimited consumption and always chasing after distant markets, with destructive consequences for local economies, communities, and all living things.

This is the potential of Slow Money: to begin to reorient capital away from endless cycles of consumption and a relentless focus on markets, towards a new economy that is focused on quality and human relationships, on our relationships to one another and to the land. After all, what is at the base of the economy? At the base of the economy is soil fertility. If we use money like synthetic fertilizer, we will get artificial growth, which can only last for awhile, but which lacks sustaining relationships with the earth. If we use money like manure, we may have a chance to create an economy built on lasting, healthy relationships. We may create a new breed of investors who refuse to accept unnatural returns.

There will be those who will be tempted to say, "But this is not economics. This is something else. This is New Age-y talk that is not a serious response to the enormity of the problems that face us!" I could not disagree more. This is the new economics, the economics of harmony and quality, the economics of honesty and patience and health, upon which our future depends.

This book is asking the most fundamental of questions, and, in this era of great complexity and financial sophistication, we need the courage to ask the most basic of questions. We need these questions to be asked by those who have the experience and vision to begin to answer them. If we cannot discover ways to invest as if food, farmers and fertility matter, how long can we expect our false progress to last? In the answers to such questions lies our journey to a new economy and a new culture.

Carlo Petrini
Founder, Slow Food International
Bra, Italy
August, 2008

prologue

"Mr. Gandhi," a reporter asked during Gandhi's 1930 visit to England, "What do you think of Western civilization?"

"I think it would be a very good idea," he replied.

I did not want to start this book with a reference to Gandhi. I really and truly did not. Having been in and around the sustainability movement for a few decades now, I've heard enough Margaret Mead and Gandhi and Einstein quotes to last several lifetimes.

But if Tom Robbins can start a novel with a reference to a beet, then anything goes.[1]

I am making light, because the subjects at hand are ponderous, because I have an inveterate skepticism when it comes to the lifeless and humorless prognostications of the experts, and because I have had it up to—*here!*—with the opaque, impersonal, deathly dull dismalness of the dismal science of economics. In our search for answers to the great questions of our day, we would be well served to listen less to economists and more to philosophers, poets, ecologists, entrepreneurs, and farmers.

To see what might lie beyond the Era of Economics, we must look above the top line and below the bottom line. I mean this almost literally. Above the top line is the region of the "meta," what

---

1. The opening line of Tom Robbins' *Jitterbug Perfume* is "The beet is the most intense of vegetables."

E. F. Schumacher called *meta-economics*. Below the bottom line is the territory of the "sub," as in *subterranean*, not in the sense of journeying to the center of the earth or anything that science-fictional, but something equally fantastic and preposterously too non-commodifiably invisible to the modern and postmodern mind: the rich, symbiotically phenomenal, mysteriously fertile life of the soil.

"A very good idea" would be a civilization that did not strip its topsoil, turn it into cheap food and highly processed food products of questionable nutritional value, and put its faith in markets at the expense of places.

Civilization is a big idea. So is the idea that as soil goes, so goes civilization. So is the idea that as money goes, so goes the soil.

We don't need any more big ideas. We need small ideas. Beautiful ideas. Beautiful because they lead to a large number of beautiful, small actions, the kind alluded to by Wendell Berry: "Soil is not usually lost in slabs or heaps of magnificent tonnage. It is lost a little at a time over millions of acres by careless acts of millions of people. It cannot be solved by heroic feats of gigantic technology, but only by millions of small acts and restraints."[2]

So, this book is small, in the hope that it may be beautiful. No grand theories or complicated metrics. Rather, intimations regarding a new way of thinking about the connections between food and soil and capital, arising from the experiences of someone who has been working for thirty years along the boundaries of venture capital and philanthropy, observing firsthand the disorienting

2. Wendell Berry, "Conservation and the Local Economy," in *Sex, Economy, Freedom, and Community* (New York: Pantheon, 1993).

impacts of what leading venture capitalist John Doerr has called "the greatest legal accumulation of wealth in history."

There is another kind of erosion at work, just as surely, here: erosion of social capital, erosion of community, erosion of an understanding of our place in the scheme of things.

It takes roughly a millennium to build an inch or two of soil; it takes less than forty years, on average, to strip an inch of soil by farming in ways that are more focused on current yield than on sustaining fertility. A third of America's topsoil has eroded since 1776. In the 1970s, the United States lost four billion tons of soil per year. Roughly a third of all farmland in the world has been degraded since World War II, with annual soil erosion worldwide equivalent to the loss of 12 million hectares of arable land, or 1 percent of total arable land. About a third of China's 130 million hectares of farmland is seriously eroded, and Chinese crop yields fell by more than 10 percent from 1999 to 2003, despite increasing application of synthetic fertilizers.[3]

Awareness of the centrality of soil health is nothing new. Aristotle laid the foundation for the humus theory of plant nutrition; his student, Theophrastus, is often called "the father of botany." The *homo* of *Homo sapiens* is derived from the Latin, *humus*, for living soil. Leonardo da Vinci observed, "We know more about the movement of the celestial bodies than about the soil under foot." Darwin spent the last years of his life studying the role of earthworms in soil fertility. After World War I, Sir Albert Howard, perhaps the father of twentieth-century organic agricul-

3. David Montgomery, *Dirt: The Erosion of Civilizations* (Berkeley: University of California Press, 2007).

ture, heralded the problems that would follow the manufacture of synthetic fertilizers by munitions factories looking for new postwar markets for nitrates: Fertilizers offered farmers boosts in yield but had deleterious effects on the health of microorganisms and the processes of growth and decay that are vital to the preservation of humus. In the first decade of the twenty-first century, despite beyond-explosive growth in our knowledge of everything from atomic energy to galactic motion, our ignorance with respect to life teeming in the soil remains humbling: It is estimated that in a gram of soil, there are billions of single-celled organisms and millions more multicelled ones, as well as over four thousand species, most of them not yet named or studied by scientists.[4]

Yet we have slipped during the past half century, as if pulled by the gravitational or centripetal forces of population growth, technological innovation, consumerism, and free markets, into a food system that treats the soil as if it were nothing more than a medium for holding plant roots so that they can be force-fed a chemical diet.

We have become dependent on technology and synthetic inputs, subsidized by what was, until very recently, cheap oil, which facilitated not only the production of nitrogen fertilizer, but also the management of large-scale, mechanized farms and the energy-intensive system of processing and long-range transportation necessary to bring agricultural products to distant markets. Agriculture accounts for more than 20 percent of U.S. greenhouse gas emissions—all the more shocking when one realizes that recent science indicates that fertile soil is a potent carbon sink, holding the potential to play a significant role in remediating global warming.

---

4. David Suzuki, "How Sustainable Is Our Development?" April 2, 2003.

The problems of our food and agricultural systems go beyond Peak Oil and Peak Soil, however. Aquifer depletion, biodiversity decline, widespread use of pesticides and other toxics, industrial feedlots that pose health and waste-management problems, nutrition and food safety challenges that attend centralized processing, the decline of rural economies, price volatility in global commodities markets:[5] It is quite a litany, surprising in its breadth and even more surprising in the degree of its invisibility when seen through the lens of the modern economy.

You wouldn't use a 747 to go to the corner store for a quart of milk. You wouldn't use a backhoe to plant a garlic bulb. You wouldn't use a factory to raise a pig. You wouldn't spray poison on your food. You wouldn't trade fresh food from family farms down the road for irradiated or contaminated or chemical-laden or weeks-old food from industrial farms halfway around the world. You wouldn't create financial incentives for farms to become so large that they need GPS technology to apply chemical inputs with quasi-military precision. You wouldn't design a system that gets only nine cents of every food dollar to the farmer. You wouldn't allow topsoil to wash down the Mississippi River, replete with pesticides and fertilizer residues, creating a dead zone the size of Rhode Island in the Gulf of Mexico. You

5. In the first few months of 2008, bread prices in Afghanistan increased 60 percent. In a single day in February, global wheat prices shot up 25 percent when the government of Kazakhstan announced it might curb exports. In 2007, the price of soybean oil in Bangladesh increased 60 percent. The FAO's global food index rose 40 percent in 2007, to the highest level on record. The impact of biofuels? Burgeoning Chinese demand? Increasing global meat consumption? Peak oil? Harvests disrupted by droughts and "freak weather"? Time magazine reports that the answer is: All of the above. (*Time*, February 27, 2008).

wouldn't use fifty-seven calories of petro-energy to produce one calorie of food energy.[6]

No, no one ever sat down and designed such a system. Yet it is precisely such a technology-heavy, extractive, intermediation-laden food system that we now need to remediate and reform.

This is the system that has evolved in the wake of global capital markets and the investors who use them, much as industrial farmers use their land—as a medium into which to pour capital in order to harvest maximum yield.

In August 2007, at the Twenty-Fifth Anniversary Gala for the Rocky Mountain Institute, eminent panelists tried to answer questions posed by moderator Thomas Friedman: "If this is a win-win-win, if these new technologies and design solutions are so elegant and so profitable and so clean, what is holding them back? Where is the resistance to these innovations coming from?" Unexpectedly, since this was not a finance conference, the group discussion zeroed in on CEO compensation, short-term financial incentives, and the structure of capital markets.

Inventor Dean Kamen opined from the dais: "Venture capitalists have great enthusiasm but short attention spans. We are stuck in a nineteenth-century way of thinking that leads to large-scale, centralized production and power generation. We don't have the

6. "A one-pound box of prewashed lettuce contains 80 calories of food energy. According to Cornell ecologist David Pimental, growing, chilling, washing, packaging, and transporting that box of organic salad to a plate on the East Coast takes more than 4,600 calories of fossil fuel energy, or 57 calories of fossil fuel energy for every calorie of food. (These figures would be about 4 percent higher if the salad were grown conventionally.)" Michael Pollan, *The Omnivore's Dilemma* (New York: Penguin Press, 2006).

mind-set to really invest for the long-term in small-scale solutions that would improve life for billions of people."

Such questions and observations lead to the premise for a new kind of financial intermediation, going by the improbable name of "slow money."

That premise is this: The problems we face with respect to soil fertility, biodiversity, food quality, and local economies are not primarily problems of technology. They are problems of finance. In a financial system organized to optimize the efficient use of capital, we should not be surprised to end up with cheapened food, millions of acres of GMO corn, billions of food miles, dying Main Streets, kids who think food comes from supermarkets, and obesity epidemics side by side with persistent hunger.

Speed is a big part of the problem. We are extracting generations worth of vitality from our land and our communities. We are acting as if the biological and the agrarian can be indefinitely subjugated to the technological and the industrial without significant consequence. We are, as the colloquial saying puts it, beginning to believe our own bullshit.

Which reminds me of a story.

About fifteen years ago, I was turning a horse stall into my office. My first project was to shovel out the dried horse manure and shovel in sand, in advance of the construction of a wooden floor.

One day, reflecting on the transition from equine to intellectual, I realized, "How appropriate: from horseshit to bullshit."

No discussion of the disconnect between capital markets and the land is complete without at least one reference to manure.

* * * * *

Let's throw in a few bees and pigs, too:

> The story of Colony Collapse Disorder and the story of drug-resistant staph are also the same story: Both are parables about the precariousness of monocultures. Whenever we try to rearrange natural systems along the lines of a machine or a factory, whether by raising too many pigs in one place or too many almond trees, whatever we may gain in industrial efficiency, we sacrifice in biological resilience. The question is not whether systems this brittle will break down, but when and how, and whether when they do, we'll be prepared to treat the whole idea of sustainability as something more than a nice word.[7]

Michael Pollan's reference to parable is telling. Just as easily, he could have referred to myth.

We are quick to assume that no battle between myths, or no myth at all, could hold sway over the modern mind. Yet could it be called anything other than myth, the story that is powerful enough to have us believing that unlimited economic growth is not only possible but desirable, despite the rapidly accumulating data to the contrary? What else but a myth could be powerful enough to convince us that what made sense as an economic organizing principle in a 1-billion-person planet with a $2 trillion global economy would still be appropriate in a 6.6-billion-person planet with a $66 trillion global economy? What else but a myth could make the soil seem expendable? What else but a myth could be powerful enough to convince us that there is no such thing as

---

7. Michael Pollan, "Our Decrepit Food Factories," *New York Times Magazine*, December 16, 2007.

a company that is too big, intermediation that is too complex, or money that is too fast?

There *is* such a thing as money that is too fast.

Money that is too fast is money that has become so detached from people, place, and the activities that it is financing that not even the experts understand it fully. Money that is too fast makes it impossible to say whether the world economy is going through a correction in the credit markets, triggered by the subprime mortgage crisis, or whether we are teetering on the edge of something much deeper and more challenging, tied to petrodollars, derivatives, hedge funds, futures, arbitrage, and a byzantine hyper-securitized system of intermediation that no quant, no program trader, no speculator, no investment bank CEO, can any longer fully understand or manage. Just as no one can say precisely where the meat in a hamburger comes from (it may contain meat from as many as hundreds of animals), no one can say where the money in this or that security has come from, where it is going, what is behind it, whether—if it were to be "stopped" and, like a hot potato, held by someone for more than a few instants—it represents any intrinsic or real value. Money that is too fast creates an environment in which, when questioned by the press about the outcome of the credit crisis, former treasury secretary Robert Rubin can only respond, "No one knows."

This kind of befuddlement is what arises when the relationships among capital, community, and bioregion are broken:

> There is an appropriate velocity for water set by geology, soils, vegetation, and ecological relationships in a given

> landscape. There is an appropriate velocity for money that corresponds to long-term needs of communities rooted in particular places and to the necessity of preserving ecological capital. There is an appropriate velocity for information, set by the assimilative capacity of the mind and by the collective learning rate of communities and entire societies. Having exceeded the speed limits, we are vulnerable to ecological degradation, economic arrangements that are unjust and unsustainable, and, in the face of great and complex problems, to befuddlement that comes with information overload.[8]

As long as money accelerates around the planet, divorced from where we live, our befuddlement will continue. As long as the way we invest is divorced from how we live and how we consume, our befuddlement will worsen. As long as the way we invest uproots companies, putting them in the hands of a broad, shallow pool of absentee shareholders whose primary goal is the endless growth of their financial capital, our befuddlement at the depletion of our social and natural capital will only deepen.

On a societal level, momentum toward an incipient epiphany about money, enterprise, and nature has been building for a few decades.

Perhaps the poster child for this process is Ray Anderson, whose self-described epiphany upon reading Paul Hawken's *The Ecology of Commerce* has reached the status of legend in the field of sustain-

8. David Orr, as cited in *Our Land, Our Selves*, edited by Peter Forbes (San Francisco: Trust for Public Land, 1999).

able business. Anderson's "spear in the chest" impelled him, at the age of sixty and after decades at the helm of his mature manufacturing company, to move from "the way of the plunderer" to the way of the sustainability pioneer, systematically working to reduce the ecological footprint of his company. Ray has been out front, but he has certainly not been alone.

Al Gore and Muhammad Yunus have brought climate change and social enterprise to the fore. Paul Hawken has evoked the "blessed unrest" of one hundred million citizens around the world working to promote sustainability. The once-wild eco-efficiency ideas of Amory Lovins are increasingly recognized. Jenine Benyus, Bill McDonough, and others are popularizing biomimicry, design ideas from nature. Not as much in the limelight, but equally important, the birth and maturation of such organizations as Business for Social Responsibility, CERES, Social Venture Network, Investors' Circle, the Social Enterprise Alliance, the Business Alliance for Local Living Economies, Ethical Markets, and a host of for-profit and nonprofit community-development financial intermediaries have been broadening traditional notions about the responsibility of business and the nature of private enterprise. Assets under management by social investment funds have grown dramatically. Fifty thousand people attend the Natural Products Expos East and West each year to see the latest in organics and natural products. Entrepreneurial companies such as TerraCycle and Stonyfield Farm and Farmers Diner and Dancing Deer Baking Company and United Villages and Energia Global and Sun Edison and IceStone and thousands more are pioneering businesses that create not only shareholder value, but also public benefit.

With all these developments and the growing momentum of such "blessed unrest," and with the crucible of environmental and economic crises ratcheting up the immediacy of these concerns every day, why hasn't the epiphany fully epiphanized?

My answer is: *The Other Hundred Million*.

That is, while one hundred million citizen activists are promoting sustainability, another one hundred million individuals—investors and the intermediaries who represent them—are stubbornly and intently affixed to their computer screens, maximizing the growth of financial portfolios and the speed of capital, promoting a culture of moneytheism and short-term thinking, maximizing circulation, minimizing percolation, diverting irrigation from the seeds of sustainability.

This book tracks what has become, for me, a kind of long, slowly unfolding epiphany that hasn't quite fully happened yet, but in whose advent I am a firm believer.

My not-quite-yet epiphany has evolved over the years with the help of many folks, whose experience and perspectives have greatly influenced my own:

- "*Even if it does work, and I think it probably would*"—responded venture capitalist Jim Fordyce in 1983 after I had presented the case for a $500,000 investment in a consumer medical database with the potential to grow to $15 million in sales with $2 million in net income—"*it won't be big enough.*"
- "*I have a moral obligation to my investors to minimize risk and maximize returns,*" said Tony Hoberman, venture capital gatekeeper at Alliance Capital, after I had made the case in 1989 for a new venture fund

that would integrate social and environmental concerns.

- *"Does it bother anyone else that we are investing in Monsanto while we are giving grants to sustainable agriculture groups?"* Edie Muma asked in the early 1990s, launching the Jessie Smith Noyes Foundation on its path of "reducing the dissonance" between the management of its assets and its charitable purpose.
- *"I want to live in a community that has ten companies with 100 employees each,"* reflected Investors' Circle member Mark Cliggett in 1999, *"but I want to invest in one company that has 1,000 employees."*

The final ingredients in my epiphanic recipe were not added until my visit to Bra, Italy, in 2000. I stood with Cinzia Scaffidi, a member of the leadership team of Slow Food, that rather remarkable NGO that had grown since its inception at a protest against McDonald's opening in Rome in 1989 into an international community with tens of thousands of members in scores of countries, in front of her favorite cheesemonger at the town's weekly farmers market. While she was talking to him, I noticed an old photograph standing on a table.

"Cinzia, who's in that picture? Please ask him."

"That's him," she translated his reply, "standing with his parents in this same spot, when he was eight years old."

As he was in his sixties, I guessed, at that moment, this meant that his family had been selling cheese every Friday in this same place for more than fifty years, and who knows how much longer. Forget the fact that his refrigerated cheese shop on wheels would have rivaled the best cheese shop in the United States. Forget the fact that under the arches at this end of the Piazza XX Settembre were several other cheesemongers who seemed equally accustomed to their spots in the market. I shuddered when I heard the answer, suffused as I was not only with the visual and olfactory richness of the moment but also with a sense of tradition, of continuity, of culture.

When I related this upon my return, one of my friends responded, "Woody, we have a name for that in this country."

"Really? What is it?" I asked.

"Socialism."

But be forewarned: Slow money is no "ism."

Slow money is not springing full-blown from the head of an economist. Rather, it is springing from myriad small actions taken by farmers, consumers, entrepreneurs, and investors who are asking questions that need to be asked, who are responding to questions that can no longer be answered adequately by the formulas of agricultural economics that lead to $400,000 combines and hog confinement units and low-flying Grumman AgCats spraying Fiprinol.

One such question has been particularly vexing for organic farmers and organic food entrepreneurs for quite a while now, particularly as Whole Foods broke into the Fortune 500: *Is organics a movement or an industry?*

To hear farmer Joel Salatin tell it, organics is part of an historical struggle that began with the Conquistadors and is reaching its zenith with the Monsantos. "Don't you ever forget," he exhorted the organic farmers gathered at the EcoFarm conference a few years back, after they had been addressed by an executive from Whole

Foods. "You are not an industry. Don't let them try to confuse you. No, you are not an industry. You are a movement. You are part of a movement that began when the first indigenous peoples fought back against the Conquistadors. You are fighting back against the modern conquistadors, the multinational corporations, those who would patent and genetically modify life and destroy diversity."

And we may hear, the moment these words fade, Stonyfield Farm CEO Gary Hirshberg's equally weighty admonition to the contrary: "Organics is not a movement. I hate the 'M' word. We are an industry. We must be an industry. We must build industrial-scale enterprises, if we are going to get toxics out of the food chain, and in one generation."

Movement vs. industry; marketplace vs. community; Small Is Beautiful vs. Wall Street; Whole Foods vs. Slow Food. There is a struggle for our hearts and minds at the inception of the twenty-first century.

My not-quite-yet epiphany is tied, now, to the not-quite-yet Slow Money, a new, nonprofit intermediary dedicated to catalyzing the flow of capital to enterprises that support soil fertility and local food communities. The not-quite-yet part is, of course, the hard part. It is the boundary-busting part, the part that resists choosing between movement and industry, the part that insists on finding a third way, or what some are calling a "fourth sector" way.

We are in the midst of the process of invention, and one might argue that this book should have awaited the outcome. Rather, it is my hope that this book will serve to broaden our conversation and improve the outcome.

For a broad view is needed, coupled with a certain fearlessness when it comes to resisting definition by the benchmarks of industrial finance. Is this "venture capital minus" or "philanthropy plus?" What return is enough to compensate for the risks entailed in building local food systems? What return is enough to compensate for the risks entailed in *not* building local food systems? Holding such ambiguity at bay long enough to attempt radical new design work requires the kind of historical perspective and radical humanism described by Slow Food founder Carlo Petrini:

> Think about the period of the Renaissance, the period of the Enlightenment, when investors would invest in things of beauty. They left things to the world that were quite beautiful. Producers were, for the most part, artisans, and they worked in a small-scale type of economy, in a local economy. Therefore, consumption sustained agriculture of one's own region. I think we have to look more closely at that model. We do, though, have to take into account the period in which we live; we're not trying to go back in time. But this can help us in our pursuit of solutions that can improve life in this global era. We mustn't think that the current system is absolute; in fact, it has to change, and it will change.

And this change will return us, armed with global knowledge and the power of free markets, to rediscover the value of the slow, the small, and the local.

Do not be surprised to find, in these pages, a strong interest in language, and, even, in the literary. What could be more important to our humanity in this age of the Mega and the Nano, in this age of "doing the numbers," than a renewed attention to words, to the values they denote and the possibilities they connote?

Wallace Stevens, one of the twentieth century's great American poets, wrote, "Money is a kind of poetry." Perhaps much of what is contained in this book could be considered footnotes to a pecuniary poem that has not yet been written.

In order to consider such possibilities, we must be unafraid not only to develop new ways of measuring, but also new ways of thinking and new language.

It is my hope, and the hope of scores of folks who have been part of this incubation process over several years, that slow money, the set of emerging ideas, and Slow Money, the emerging NGO, will not only catalyze new sources of capital in support of local food systems and soil fertility, but will also contribute to a broader inquiry into our assumptions about fiduciary responsibility and the nature of fiscal prudence.

The material in this book was written over several years, in tandem with conferences, workshops, strategic-planning retreats, and an on-going dialogue with a wide range of stakeholders: investors, entrepreneurs, grant makers, social investment advisors, and farmers. It does not present a cohesive, linear argument or history. Quite the contrary: It presents pieces of an emerging mosaic, the gestalt that is arising from these efforts.

In January 2008, sixteen individuals gathered for three days at Sundance Village, Utah, for the founding retreat of Slow Money. As of this writing, 55 individuals have become founding, contributing members of the Slow Money Alliance (SMALL), and the

Blue Moon Fund of Charlottesville, Virginia, has provided seed support. Substantial support also has been committed by the Investors' Circle Foundation. We are currently modeling a slow-money portfolio based on fieldwork in northern Vermont, and will be using this model to foment discussion and analysis at three Slow Money Institutes, in the fall of 2008, in Vermont, Appalachia, and Northern California. The results of this work will inform decisions regarding the formation of a first Slow Money fund in 2009.

The book is organized as follows: Part One, called "Slow Money," begins with a short introductory essay of the same name. "Reconnoitering," written after this year's founding Slow Money retreat and the fieldwork in Vermont, presents the thinking behind Slow Money, both philosophical and practical, laying out the relationships among small food enterprises, soil fertility, and capital markets, the position of slow money in the history of social investing, and initial reconnaissance regarding the design of slow-money investment funds. "Back Down to Earth" is adapted from the vision-mission-strategy paper that served as background reading for the founding Slow Money retreat.

Part Two, called "Ground Zero," comprises three essays that explore, in a personal voice, the relationships among food, violence, culture, and capital. "In the Shadow of the Twin Towers" presents a thought experiment about the possibility of an alternate stock exchange. "The War on *Terroir*" explores our relation to the land and the question of nonviolence. "The Pursuit of Zero" considers the idea of food as Ground Zero in the struggle to reduce the dissonance between our ideals and our actions.

Please keep in mind that this material was written for specific audiences at specific moments in time, part of not only an evolving way of thinking, but also an evolving dialogue. As with any dialogue, a certain amount of rumination and reiteration was inevitable and essential. Hundreds of individuals participating in scores of workshops during a decade or so have been part of this process of incubation. You will therefore note, on occasion, a repeated reference to a pivotal citation or conversation, as certain insights became embedded in the DNA of this new undertaking.

What this collection of essays lacks in coherent narrative and exposition, I hope it will compensate for with transparency and immediacy, offering a real-time peek under the hood at a process of fiduciary activism, and serving as a provocation, a catalyst, an invitation.

It goes by many names: The Theory of Unintended Consequences; Too Much of a Good Thing; One Step Forward, Two Steps Back; The Tortoise and the Hare; Zeno's Paradox; The Myth of Icarus.

Rod Serling called it "To Serve Man."

Around the time that McDonald's was serving its 200 millionth burger—that is, in 1962, when I was eleven years old—I saw an episode of Rod Serling's TV series *The Twilight Zone*. In it, aliens come to earth and perform a series of miracles: curing cancer and turning the Sahara into a wheat field. They offer mankind a book entitled *To Serve Man*, but the text is in their native language. Linguists and cryptographers have at it, keen to decipher and disseminate the knowledge of this advanced civilization. Meanwhile, people eagerly accept the aliens' invitation to visit their planet.

At the end of the episode, as one of the cryptographers is boarding a spacecraft, his assistant rushes toward him, shouting: "Wait! Don't go! We've broken the code! *To Serve Man*: It's a cookbook!"

No economist ever argued more persuasively that every benefit has its cost, that there is a shadow side to the technological innovations and dramatic increases in standard of living that have characterized the second half of the twentieth century for a few billion *Homo sapiens*.

Recognizing full well the risks of invoking the likes of Gandhi, Tom Robbins, and Rod Serling in the same introductory bundle, I offer this: It is time to do more than just jog our thinking.

We must reexamine the story of progress, being honest about where we as a species seem to be going awry—if one defines nuclear weapons, 9/11, Twinkies, the continuing promulgation of a culture of rampant consumerism in the face of 6.6 billion people, climate change, volatility in global food prices, and widening wealth inequalities as "going awry." We must recognize the potentially disastrous consequences of doing the same thing over and over again—going more and more global, bigger and bigger, faster and faster—hoping for a different outcome.

Then, and only then, we may find the courage to slow money down—not all of it, of course, but meaningful quantities of it. In a world of trillions of dollars a day, this means billions of dollars a day . . . *wait* . . . did I say *billions of dollars a day*? I did, and before *The Twilight Zone* theme song starts ringing in your ears, I add: The Good Lord, and his better half, Great Gaia, did not hand us today's global financial markets on a platter, blessed and preordained. No, we invented them, and we have it in our power to reinvent them, to design what comes after them. What seems preposterous when viewed through the wrong end of the fast-money telescope seems wonderfully within reach when looked at through the lens of slow money, allowing us to set about the work of rebuilding healthy relationships among enterprises and communities and bioregions, and between investors and the enterprises in which they invest.

Let us set about this work, then, so that whoever it is that comes onto the world stage after *Homo economista* and *Homo consumerista* and the Invisible Hand of the marketplace, may, as they exit stage left, come—fully and fearlessly and wearing whiffs of humus and manure like badges of honor—into view.

PART ONE

slow money

ONE

slow money

Every 200 years or so, it seems, we arrive at a threshold moment in the history of capital and culture.

In 1600, two men in Amsterdam stood on a bridge over a canal, designing the joint stock company, minimizing risk to capital, and galvanizing the flow of investment in exploration, conquest, and export. The outlines of the New World were as yet undefined and the notion of limits to growth unimaginable.

In 1800, two men in New Amsterdam stood under a tree on the cow path that would become Wall Street, designing a stock exchange that would create hitherto unknown degrees of financial liquidity and, so, galvanize the flow of capital in support of exploration, extraction, and manufacture. Corporations were small, continents were large, industrialization was incipient, the Prudent Man and the Invisible Hand about to enjoy their considerable time in the sun, and the notion that the resilience of natural systems had limits was about to suggest itself—but only briefly, and only to be swiftly discredited and debunked.

In 2000, we are entering a period of urgent postindustrial, post-Malthusian reassessment and reconnoitering. We find ourselves on a new threshold, signals of systemic unsustainability proliferating alongside those of ever accelerating capital markets and technological innovation. Consumerism and global markets are ascendant, carbon sinks are overloaded, and the idea of limits to growth calls for radical reconsideration.

It falls to us to undertake a new project of system design: the creation of new forms of intermediation that catalyze the transition from a commerce of extraction and consumption to a commerce of preservation and restoration.

Speed is one of the defining characteristics of our age. As much as in the Age of Ones and Zeros or the Age of Gigabytes and Megatons, we live in the Age of Hockey Sticks.

Breaking the six-billion-person and the billions-of-instructions-per-second and the billions-of-shares-per-day barriers, we are disoriented by the seductions of speed. At the same time, our knowledge of the world becomes considerably more complete,

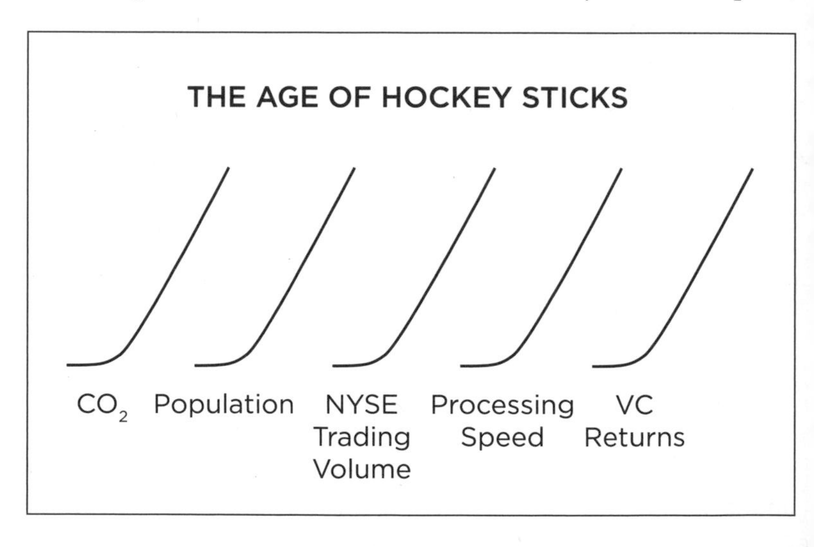

affording us truer perspectives on the incompleteness of economics disconnected from bioregions and communities, markets disconnected from places, wealth disconnected from health.

GDP growth driven by subdivisions and highways and Mustangs and 727s made sense in a pre-smog, pre-urban-blight, pre-sprawl world. Buy Low/Sell High made sense in a world that could not conceive of a $350 billion Wal-Mart, a $53 million Christmas bonus, a $400 million golden parachute, or a China that is building one coal-fired power plant a week and more roads in 2008 than it had in the preceding fifty years.

Now, a thousand billionaires and a billion thousandaires signal structural limits to the power of industrial finance. Close to four hundred parts per million of carbon in the atmosphere signals limits to the economics of maximum growth. The campaign to drag 1 or 2 percent of foundation assets back across the iron curtain between asset management and grant making signals the limits of the culture of Wealth Now/Philanthropy Later. And, most recently, the subprime mortgage collapse signals the limits of ever accelerating, ever more complex, derivative-driven financial markets.

Organized from "markets down" rather than from "the ground up," industrial finance is inherently limited in its ability to nurture the long-term health of community and bioregion. These limits are nowhere more apparent than in the food sector, where financial strategies optimizing the efficient use of capital have resulted in cheap chemical-laden food, depleted and eutrophied aquifers, millions of acres of GMO corn, trillions of food miles, widespread degradation of soil fertility, a dead zone in the Gulf of Mexico, and obesity epidemics side by side with persistent hunger.

"Food," as the poet Gary Snyder observed, "is the field in which we daily explore our harming of the world." It is also the field in which we daily explore the boundaries between investing and philanthropy. Using global markets as our guide, we choose commodity production over soil fertility, leaving the vast majority of sustainable-agriculture enterprises with little or no access to either investment capital or philanthropic support.

While organics grows at 20 percent per annum on its way past 3 percent of U.S. food-industry revenues, and Whole Foods' emergence as a Fortune 500 company heralds wider market acceptance of organics, the facts on the ground remain stark. Only 0.5 percent of U.S. farmland is organic. Only 0.1 percent—less than $50 million per annum—of U.S. foundation grants go to sustainable agriculture. Only 0.1 percent of the USDA's budget supports organics. Roughly the same order of magnitude of venture capital targets organics, and most of that goes to organic brands that have limited relevance to the health of local food systems.

The challenge of reintegrating social and environmental concerns into the financing of food mirrors similar processes underway in broader capital markets and philanthropic arenas.

Socially responsible investing, mission-related and program-related investing by foundations, venture philanthropy, social entrepreneurship, local economies, consumer demand for organics and green products—these are the first stages of a more profound fiduciary realignment. Some of these initiatives remain incremental and ambiguity-laden. Others are indicators of more fundamental, tectonic shifts along the boundaries of for-profit and nonprofit, shareholder and stakeholder, global investor and local citizen.

This process of economic and cultural transformation calls for a new prudence, a new urgency, a new vision of capital markets

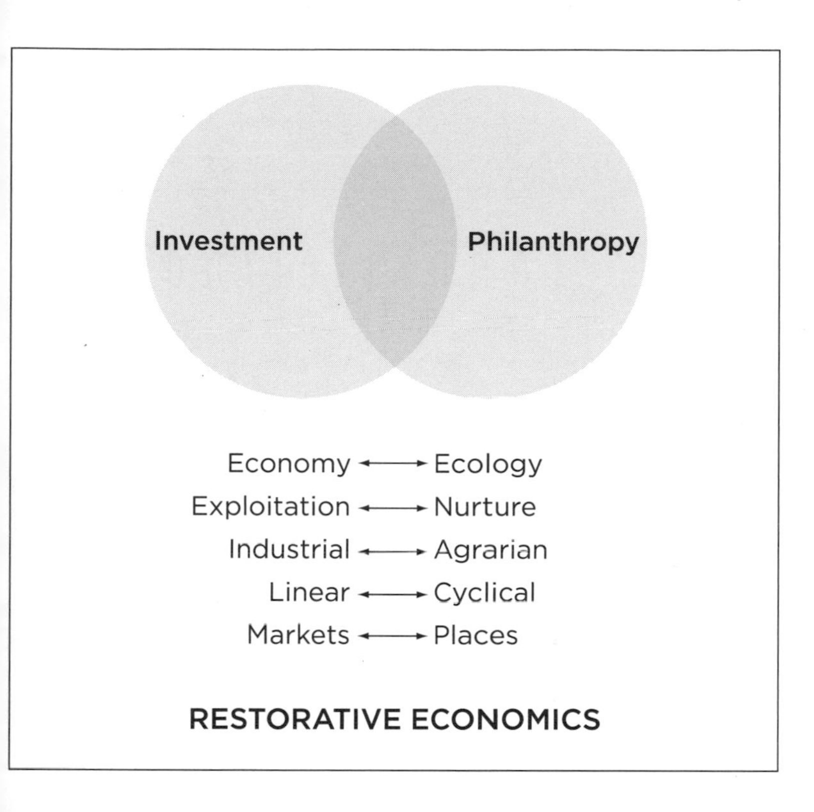

designed to usher in the age of restorative economics, integrating into the theory of fiduciary responsibility and the practice of asset management principles of carrying capacity, care of the commons, sense of place, cultural and biological diversity, and nonviolence.

One of the principal measures of our success will be the extent to which we have catalyzed substantial new capital flows to enterprises that create economic opportunity while respecting, protecting, and promoting the fertility of the soil.

TWO

reconnoitering

It's as if we have been so busy answering Question One on the Terra Madre[1] of All Final Exams that we overlooked the fact that there are two questions.

Question One is this: *Can we invent eco-efficiency and bioremediation technologies that will prevent irreversible damage from climate change?*

The answer is unfolding. Billions of dollars of venture capital are invested annually in solar and wind and biofuels, cellulosic ethanol, fuel cells, ultra-efficient lighting, carbon sequestering "mechanical trees" that can pull carbon out of the atmosphere in globally significant quantities, and genetically engineered bugs to eat oil spills and toxics. The search is on for technologies that will enable us to maintain our way of life and grow the economy, while dramatically reducing our ecological footprint.

Unfortunately, even the best answer to Question One is insufficient for a passing grade.

---

1. Terra Madre is a network of 1,600 food communities from 150 countries, organized by Slow Food. Every two years, the Terra Madre conference brings together 5,000 small farmers, as well as 1,000 chefs and academics, in support of sustainability, indigenous culture, and social justice.

We also need to answer Question Two: *If we cannot grow the economy without destroying the fertility of the soil, how can we, no matter how clean our machines, hope to thrive, or, even, survive?*

Every era has had its neo-Malthusian muckrakers and malcontents, on the one hand, and its Armageddon-prognosticating fundamentalists, on the other.

We baby boomers will be forgiven our particular penchant for big booms and big bangs. We were raised in the shadow of World War II's big nuclear booms (or were they bangs?). The possibility of going out with a bang was very much in the fore in those early decades of nuclear proliferation. We muddled through. Nuclear war was never waged. We landed on the moon. The Berlin Wall came down. McDonald's blew through billions of burgers. DDT and dioxins were outed. CFCs were banned. The ozone hole seemed to stay pretty much to itself, over Antarctica.

The sky did not fall.

The earth, however, was slipping away. One petrochemically farmed field at a time. Tons of topsoil at a time. Trillions of microorganisms at a time.

It never occurred to us that it might not be a bang, but a whole bunch of whimpers that would do us in: too many cars and too many smokestacks and too many suburbs and too many malls and too many inner cities and too many toxins and too many sneakers made in too many sweatshops and too many mortgages and too many hamburgers and too many antibiotics and too much N-P-K fertilizer and too much pesticide and herbicide and too many rural communities collapsing and too many small and midsize farmers giving up the ghost.

All of these whimpers were drowned out by the roar of the stock market. The Dow Jones Industrial Average was 150 in 1945; by 2007, it had passed 14,000.

If there had been a manual of civilization way back when, it might have read: Never start a ten-thousand-year epoch of agriculture-dependent civilization without first understanding soil fertility, biodiversity, carrying capacity, and the relationship between economics and ecology.

Which is to say: We could not, in those early days of Mesopotamia, or even much later, in the early days of the steam engine and the joint stock corporation, have anticipated the limits of agriculture or the limits of economics. We could not have been expected to understand, when the first clay tablets were being scribed with shepherd's inventories and the first wheat fields were being cultivated, how agriculture would pave the way from hunter-gathererdom all the way to Wall Street. Any more than Descartes could have been expected to anticipate Thoreau or Einstein. Any more than the Wright Brothers could have been expected to anticipate the bombing of Dresden or the advent of the stealth bomber. Any more than Henry Ford, exuberant with the opportunities of mechanical horsepower and the prospects of manureless cities, could have been expected to anticipate, a scant one hundred years hence, the 405 in Los Angeles during rush hour or suburban sprawl or the aggregate particulate emissions of 500 million tailpipes.

And it would have taken an oracle of Delphic capabilities to have foreseen that as the whole planet geared up and heated up and sped up in the twentieth century, responding to the triple-threat explosions of population growth, technological innovation, and financial markets, the future would hinge in such significant measure on a very different triple threat: the small, the local, and the slow.

In relation to a tree or a mountain or a river or a valley or a phalanx of armored vehicles, a person is small. In relation to the night sky and the universe of which it is a part, the earth is very, very small. In relation to history, a footnote is small. In relation to Wal-Mart, which has $350 billion in sales, Patagonia and Stonyfield Farm, each with hundreds of millions of dollars of revenues, are small. Butterworks Farm, with a $1 million yogurt business operated sustainably from one 325-acre farm and a herd of fifty cows, is very much smaller yet, but is large compared to many of its neighboring farms, for whom such revenues or branded-product success are enviable.

This is the beauty of the word *small*. You cannot use it absolutely. It only exists in relation. It forces qualitative judgment. It implies appropriate scale and challenges the intelligence to figure out just what in the world that means.

Similarly with the word *local*. Is *local* a specific number of miles from Point A to Point B? Is it a specific political or contractual designation? How does it relate to a bioregion or a watershed? In the word *local* are connotations of rapport, relationship, rootedness, and the kind of responsibility that is the reciprocal of anonymity and absentee ownership. In the word *local* is a hint of the possibility of *localization*, that which is bubbling up in the wake of globalization. The quantitative imprecision of the term does not diminish its importance. As with the word *small*, its importance lies precisely in a reliance on qualitative distinctions:

> Most of the "conspicuous developments of economics in the last quarter of a century" (referred to by Professor Phelps Brown) are in the direction of quantification, at the expense of the understanding of qualitative differences.... For example, having established by his purely quantitative methods that the Gross National Product of a country has risen by, say, five percent, the economist-turned-econometrician is unwilling, and generally unable, to face the question of whether this is to be taken as a good thing or a bad thing. He would lose all his certainties if he even entertained such a question: Growth of GNP must be a good thing, irrespective of what has grown and who, if anyone, has benefited. The idea that there could be pathological growth, unhealthy growth, disruptive or destructive growth is to him a perverse idea which must not be allowed to surface. A small minority of economists is at present beginning to question how much further "growth" will be possible, since infinite growth in a finite environment is an obvious impossibility; but even they cannot get away from the purely quantitative growth concept. Instead of insisting on *the primacy of qualitative distinctions*, they simply substitute non-growth for growth, that is to say, one emptiness for another.[2]

Which brings us around to that all-important, qualitative-distinction-laden word *slow*.

---

2. E. F. Schumacher, *Small Is Beautiful: Economics as if People Mattered* (London: Blond & Briggs, 1973).

"Money only knows one speed," the scion of one of America's wealthiest families said during a public discussion. "Money only goes fast, faster, fastest. Try to slow it down, and you'll just end up with sloppy investing."

To which someone responded, "I couldn't agree more with the first half of your statement. Left to its own devices, money will go fast, faster, fastest. It is up to us to design ways to slow it down. That would not be sloppy investing, in my book. That would be wonderfully intelligent investing."

In relation to what, in comparison to what, would slow money be . . . *slow*?

Consider this: From the beginning of human history to the year 1900, the world economy grew to $600 billion in annual output. Today, the world economy grows by this amount *every two years*; global output reached $66 trillion in 2007. Here are some of the numbers:

- $3 trillion circulate through currency markets *every day*.
- Trading volume on the New York Stock Exchange, which averaged 3 million shares per day in 1960, crossed the 100-million share mark in 1982, the billion-share mark in 1997, the 2-billion-share mark in 2001, and the 5-billion-share mark in 2007.
- Revenues of Wall Street broker-dealers rose from $20 billion in 1980, to $76 billion in 1990, to $325 billion in 2000.
- To make the list of the top twenty-five hedge-fund managers in 2002 required personal compensation of at least $30 million; in 2006, $240 million.

- A share of Berkshire Hathaway that cost $8 in the 1960s was worth over $100,000 in 2007.
- Mickey Mantle's salary for the 1968 season was $100,000; in 2007, Roger Clemens got paid $20,000 *per pitch*.
- In 1982, the world had 12 billionaires; in 2000, 298; in 2008, 1,125 with a combined net worth of $4.4 trillion.
- In the United States, the wealth gap has nearly doubled since 1980, hitting levels not seen since the 1920s. The top 0.1% of Americans collectively enjoy almost as much income as the bottom 50%, with the average income in the top group 440 times that of the bottom half.[3] In China, the wealthiest 10 percent own more than 40 percent of all private assets, and inequalities are widening.[4] On a global scale, the richest 2% own more than half of all household wealth.
- U.S. foundation assets grew from $32 billion in 1980 to $550 billion today.
- The U.S. venture capital industry exploded from an average of $3 billion invested per year in the 1980s to a peak of $100 billion invested in 2000, and then settled in the range of $20 billion per annum, with this money invested in three or four thousand high-tech companies out of the more than 500,000 new corporations started every year—that is, invested only in those with a shot at being "the next Google," growing to billions of dollars of market capitalization in a few years.

---

3. Johnston, David Coy, "Income Gap Is Widening, Data Shows." *New York Times*, March 29, 2007. www.nytimes.com/2007/03/29/business/29tax.html
4. "The Richer China Grapples with Widening Wealth Gap." *People's Daily Online*, December 18, 2006. http://english.peopledaily.com.cn/200612/18/eng20061218_333648.html

- We have passed the billions-of-computer-instructions-per-second barrier, on our way, by 2015, to the rarified atmosphere of ten tera-instructions per second.

The analogy of a rocket accelerating to reach escape velocity from Earth's gravitational field has some relevance. Venture capital targets companies that are ready to "take off." But this is only a small part of the overall acceleration in financial markets:

> Computerization mothered a new dimension of financialization. . . . Just as rocket science took men into space in the 1950s, another exotic mathematics—in this case, capital asset pricing, options theory, and price and volatility models—took finance into a hitherto unexplored galaxy of profits. Turning money into equations and digital impulses—shifting to the megabyte standard, in economic journalist's Joel Kurtzman's term—allowed it to jump time and geography, creating the transnational netherworld in which traders in New York, London, and Paris warred electronically over Belgian francs and Thai baths and global arbitrage fed on shoals of cyberdecimals.[5]

The speed of financial markets, and the short-term thinking that it breeds in CEOs and investors, is both a reflection of and a cause of a broader disorientation in the culture at large. Accelerating money and accelerating computers are one thing; far more mundane manifestations of speed have profound consequences for our perception. Even 65 miles per hour—far, far slower

5. Kevin Phillips, *Wealth and Democracy* (New York: Broadway Books, 2002).

than a snail's pace relative to cyberspeed—can disconnect us from our surroundings: "Some would argue that covering more ground exposes the speeding driver to more of what is real. But, ironically, the faster we go, the less we truly see. Speed insulates us from organic detail, and space becomes not homes, neighborhoods, and individual lives, but a disembodied medium through which we move. Though more is seen, less is observed, for the depth of our understanding is inversely proportionate to our velocity."[6]

The acceleration of information and financial markets amplifies such questions about the relationship between information and knowledge, between economics and nature. For as financial time contracts, how can we maintain a healthy relationship to natural time? The contrast is striking: natural time versus money time, seasons and generations on the one hand, fiscal-year quarters and product life cycles on the other; the time it takes water to flow through soil and aquifer versus the time it takes money to cycle through mutual fund and derivative; the time it takes to create soil and oil versus the time it takes to deplete them; currency swap and collateralized debt obligation and arbitrage versus protozoa and mycorrhizal fungus and earthworm.

Fifty trillion dollars of derivatives is quite a monument to financial ingenuity, but at what risk do we dare to outsmart an ear of corn?

Farmer Gene Logsdon observes how the imperatives of money's speed play out in his field:

6. Stephen Bertmann, *Hyperculture: The Human Cost of Speed* (Westport, CT: Praeger, 1998). Bertmann reports that CBS broadcaster Charles Kuralt once said: "The Interstate Highway System is a wonderful thing. It makes it possible to go from coast to coast without seeing anything or meeting anybody."

> No ear of corn has ever heard of 6 percent interest, much less 12 percent. To keep pace with that kind of inflation, the farmer is forced to squeeze every kernel of production he can from the soil. He increases applications of toxic chemicals known to adversely affect soil microorganisms. He buys heavier and heavier machinery, which compact his soil, so that yields drop. He applies increasing amounts of fertilizers to bring yields back up. These strategies all seem to reach points of diminishing returns. . . . Blending the opposing forces of economics and ecology into a productive farm demands supreme skill. No scientist, no ecologist, no chemist, and God knows, no bureaucrat or social scientist is going to do it. An experienced, informed farmer is the only chance.[7]

The diminishing returns of industrial agriculture run beyond the cornfield to the entire system: drastically declining numbers of "experienced, informed" farmers; increasing crop losses due to pests despite dramatic increases in pesticide use; ethical, environmental, and nutritional problems of animal confinement units and factory feedlots; collapsing fisheries; disappearing bees. Here are some of the numbers:

- In 1950, there were 25 million U.S. farmers; today, there are 2 million; 163,000 "mega-farms" produce 60 percent of the nation's food.
- In 1980, the biggest poultry processing plants in the United States handled 16 million birds a year. Today,

7. Gene Logsdon, *At Nature's Pace* (New York: Random House, 1994).

according to one industry observer, "you need 1.25 million birds a *week* just to break even."

- 230 pounds of synthetic nitrogen is applied to a typical acre of U.S. corn; as much as 50 pounds enters the surrounding environment, contributing to groundwater eutrophication.
- Despite a tenfold increase in pesticide use since 1945, crop losses due to pests have almost doubled. The National Academy of Sciences listed 440 pesticide-resistant insects in 1986 and the number is rising.
- Over the last fifty years, world irrigated area tripled. Half the world's grain crop is grown on irrigated acres, and the United Nations Food and Agriculture Organization (FAO) projects a 20 percent increase in irrigated acreage by 2030. However, aquifers are being depleted: In India, water tables are falling by up to 20 feet per year; in Northern Africa, aquifers are being depleted five times faster than they can recharge; in China, water tables have fallen by up to 300 feet, and the World Bank reports that water use exceeds sustainable flow by more than 600 million tons per year; in the United States, the Ogallala Aquifer, which supplies one in five irrigated acres, is being overdrawn by 170 million tons, or 3.1 trillion gallons, per year.
- The FAO reports that global grain yields per acre are increasing by only 1.3 percent per year, barely half the rate of 30 years ago and much more slowly than demand is growing.
- 65 percent of U.S. grain is used as livestock feed.
- Concentrated animal-feeding operations (CAFOs)

account for 40 percent of the world's meat production, up from 30 percent in 1990. China's CAFOs produce 40 times more nitrogen pollution and 3.4 times the solid waste of industrial factories. Effluent from U.S. CAFOs contaminates groundwater in 17 states. Heavy subtherapeutic use of antibiotics by livestock producers accounts for nearly half of all antibiotics used worldwide, producing new strains of antibiotics-resistant bacteria.

- The industrialization of food production—large-scale monoculture, genetically modified varieties, and the consolidation of seed production by agribusiness companies—has resulted in dramatic declines in the biodiversity of cultivated crops. In 1900, there were 7,000 varieties of apples in the United States; today, less than 1,400 remain. According to the Global Crop Diversity Trust, 95 percent of cabbage, 91 percent of corn, 94 percent of pea, and 81 percent of tomato varieties were lost during the twentieth century.[8] In less than a century, India's cultivation of 30,000 indigenous varieties of rice is shrinking to less than 100, with the top 10 expected to account for over three-quarters of the subcontinent's rice acreage.[9]
- Worldwatch and FAO estimate that 76 percent of global fish stocks are overfished, with 366 out of 1,519 fisheries worldwide collapsing.
- Beekeepers in the United States and Europe have suffered dramatic losses, some as high as 70 percent,

8. Claire Hope Cummings, *Uncertain Peril: Genetic Engineering and the Future of Seeds* (Boston: Beacon Press, 2008).
9. Vandana Shiva, *Monocultures of the Mind* (London: Zed Books, 1993).

> from what is being called "colony collapse disorder," with causes under investigation including crowding, inadequate nutrition, pesticide exposure, infection, and the combined effects of prophylactic antibiotics and miticides.[10]

There is little dispute about the numbers. The only real dispute is about what story accompanies the numbers. Are these merely the unfortunate side effects, or "externalities," that accompany economic growth? Are these problems that will be handled, each in turn, by technological fixes? Or, are these serious, structural cracks in the foundation of industrialization? Are these feedback loops calling into question the linear fallacies of economic growth?

And then there are the numbers and the stories of the soil.

Every second, a dump-truck load of topsoil is carried by America's rivers into the Caribbean; every year, hundreds of millions of tons of topsoil erode from farmers' fields in the Mississippi River basin. "An estimated 36% of the world's cropland is suffering from a decline in inherent productivity from soil erosion," reports Lester Brown, one of the world's leading ecosystem monitors. Globally, we are losing more than 10 million hectares of arable land each year, with soil loss exceeding new soil production by 23 billion tons, resulting in the loss of 0.5 percent or more of the world's soil fertility, annually.

Worldwatch Institute calls it "the quiet crisis of the world economy." Let us call it Peak Soil.

---

10. The data in this section is drawn primarily from Paul Roberts, *The End of Food* (New York: Houghton Mifflin, 2008) and Samuel Fromartz, *Organic, Inc.* (New York: Harvest Books, 2007), as well as from Worldwatch Institute, the Woodrow Wilson International Center for Scholars, and Natural Resources Defense Council.

The phrase *Peak Soil* is currently being used by anti-biofuel advocates, highlighting the problems of diverting arable land from food production to fuel production. The phrase hints at something more ominous than the current controversy about ethanol, however.

Peak Soil suggests the exhaustion of a resource that is ultimately far more vital to civilization than either oil or ethanol. "Societies that deplete natural stocks of critical renewable resources—like soil—sow the seeds of their own destruction," writes David Montgomery, professor of earth and space sciences at the University of Washington. "Technology, whether in the form of new plows or genetically engineered crops, may keep the system growing for a while, but the longer this works the more difficult it becomes to sustain—especially if soil erosion continues to exceed soil production."[11]

Beyond a small number of geologists, environmentalists, and organic farmers, the idea of Peak Soil seems as inconceivable today as global warming seemed two generations ago. One is tempted to ask with all incredulity: *Could we really run out of dirt?*

Which leads us back to Question Two in the Terra Madre of All Final Exams. The question is not only a question of erosion, it is a question of fertility. It is a question of nematode and protozoa, a question of fungi and bacteria, a question of microorganisms too numerous to count. Studying virgin prairie soils and cultivated soils in Missouri in the 1930s, soil scientist Hans Jenny found that after sixty years of cultivation, farm soils had lost a third of their organic matter, *even in the absence of soil erosion*. Our understanding of the soil has not advanced much since then.

---

11. David Montgomery, *Dirt: The Erosion of Civilizations* (Berkeley: University of California, 2007).

"Soil science is in its infancy," geneticist David Suzuki wrote in 1998. Our attention remains focused on what goes on above the ground. To paraphrase organic farmer Eliot Coleman, we are so busy feeding the plants that we have forgotten to feed the soil. (And he's talking about compost and manure, crab shells and seaweed and crop residues and peat, not empty, synthetic calories of the N-P-K kind.)

Paul Roberts summarizes the state of our knowledge about soil organic matter in his recent book, *The End of Food*:

> The real Achilles' heel of the Green Revolution was, and is, fertilizer. By conservative estimates, more than a third of the Green Revolution yield increases came directly from using more fertilizer. And yet, as American and European farmers were discovering, while fertilizers were a necessary ingredient for modern high-yield agriculture, they were not sufficient to ensure its success. Although African farmers saw massive yield increases within the first few years of adopting new techniques, in a relatively short time, something odd happened—yields fell unless farmers added steadily greater applications of nitrogen and other fertilizers. This effect was so dramatic that over the course of twenty years, a farmer would need to double his nitrogen applications simply to maintain his yields at their initial level.
>
> Why this change occurs isn't entirely clear, but research suggests that under intensive agriculture methods, soils will lose not just macronutrients—nitrogen, phosphorus, and potassium, which can be replaced synthetically—but the carbon-rich organic matter left over by decaying

> plants and animals. This organic matter is key to good crop yields. The more organic matter in the soil, the more rainwater the soil can absorb and retain, which means water for crops. Organic matter also helps the soil particles stick together, reducing the risk of wind and water erosion. As well, soils rich in organic matter have a greater capacity for additional nutrients—that is, they can absorb more fertilizers, whether natural or synthetic, and convey those nutrients more readily to other plants. In short, adding synthetic fertilizers to lands rich in organic matter—such as the American Midwest and certain parts of Africa—could indeed bring massive yield increase. The problem is that soil organic matter, or SOM, can be depleted when farmers raise too many crops without replenishing nutrients with cover crops or manure or other fertilizers. And once SOM begins to fall, the soil's capacity to hold and transport synthetic nutrients also falls. . . .[12]

Two things strike me in Roberts' summary of the role of SOM. The first is the phrase "isn't entirely clear."

Why isn't it? Entirely clear, that is. Given all that we know about genetics and chemistry and biology, why don't we understand clearly the soil mechanisms that result in diminishing returns for synthetic-fertilizer application? Have we failed to research it adequately, preferring to invest in the engineering of new plant varieties? Or is it unknowable, an irreducible mystery, a secret buried in a vast symbiotic complex? Either way, shouldn't our lack of clarity suggest a certain humility, a certain precaution, a certain prudence?

12. *The End of Food.*

The second is the absence of the word *microorganism*. Wouldn't a layman's prudence, the nonscientific, appropriately skeptical prudence of anyone who'd ever tended a few raised beds or knew the difference between a hornworm and an aphid, lead in the direction of what might prove, in the end, intuitively obvious: Applying large amounts of synthetic fertilizer, herbicides, and pesticides *must* adversely affect soil microorganisms that are essential to the processes of decay and fertility?

The word *microorganism* does not appear in *The End of Food* index, and I don't recall reading it in anywhere in the book. Roberts touches on seemingly every facet of the food system: the historical roots of agriculture, food processing, nutrition, obesity, farm subsidies, commodity markets, fast food, slow food, Whole Foods, genetic engineering, international trade, aquifer depletion, and soil erosion. Yet, his lens as an investigative journalist does not focus on microorganisms.

Enter David Suzuki, who zooms in on them:

> Soil microorganisms comprise a major portion of the total diversity of life. In this dark, teeming world, minute predators stalk their prey, tiny herbivores graze on algae, thousands of aquatic microorganisms throng on a single drop of soil water, and fungi, bacteria and viruses play out their part in this invisible stage. In their life and death these organisms create and maintain the texture and fertility of the soil; they are caretakers of the mysterious life-creating material on which they, and we, depend absolutely. . . .
>
> Despite their microscopic size, soil microorganisms are so abundant that they make up a significant biomass; in fact, they may be the major life-form in any given area. . . .

> Technologically advanced nations have not been using the soil in a sustainable way; instead, they have been "mining" the soil by removing its organic content without replacing it, thereby compromising its future productivity for the sake of the enormous harvests of today.[13]

Each gram of fertile soil contains hundreds of billions of bacteria and actinomycetes, hundreds of thousands of fungi and algae, and tens of thousands of protozoa, nematodes, and other microfauna. That was *each gram* of fertile soil. Let soil quants extrapolate this to shovelfuls, yards, tons, and acre-feet. I see your trillions, and I raise you trillions and trillions more: How many trillions of microorganisms are in the soil for every dollar of derivatives that circulate the globe?

Earthworm numbers are less stratospheric. There can be as many as two million earthworms in an acre of fertile soil. Or should we measure them in pounds per acre (estimated at from 356 to 612 by one researcher in New Zealand)?[14] Or should we measure them in burrow-miles per acre (estimated at 1,100)?[15] Or should we measure them in castings per acre (estimated at 33 to 40 tons)? Darwin estimated that 50,000 earthworms carry 18 tons of soil to the surface of an acre.

The number of earthworm species is a moving target. Edwards and Lofty estimated 1,800 species in their 1972 work, *The Biology of Earthworms*. According to encylopedia.com, the number is 2,200.

---

13. David Suzuki, *The Sacred Balance: Rediscovering Our Place in Nature* (Amherst, NY: Prometheus Books, 1998).
14. K. E. Lee, "The Role of Earthworms in New Zealand Soil," *Tuatara* 4, no. 1 (July 1951).
15. Hugh Martin, in *Ontario Farmer*, reprinted by Ecological Agricultural Projects, McGill University, 1994.

According to compost-bin.org, the number is more than 4,400, in three categories: endogeic, anecic, and epigeic. Amy Stewart reports 4,500 species (739 genera, 23 families, several suborders and superfamilies, and two orders).[16]

"There is no better soil analyst than the lowly earthworm," Sir Albert Howard wrote in his introduction to a reprint of Charles Darwin's final work, *The Formation of Vegetable Mould through the Actions of Worms with Observations on Their Habits.*

Referring to "the war in the soil" and "the battle of organic and inorganic," Howard offered that it would be the earthworm that would interpret Mother Earth's final "decision."

Howard's war in the soil harkens back to Gene Logsdon's evocation of farmers reconciling "the opposing forces of ecology and economics." These opposing forces manifest themselves as a battle between SOM (soil organic matter) and SCM (supply-chain management):

> There is a growing consensus outside the industry that the crisis may already be beyond growers' ability to fix. Despite the media's focus on the feral pigs as the killer vector in the spinach outbreak, for example, researchers have identified dozens of other "nodes of risk" where pathogens could have breached the industry's safety systems. And, as with the meat business, many of those nodes were created by the very technologies and business practices that allow the industry to deliver ever greater volumes of produce year-round at declining costs. "There guys are in supply chain

16. Amy Stewart, *The Earth Moved: On the Remarkable Achievements of Earthworms* (Chapel Hill: Algonquin Books, 2004).

> mode," says Trevor Suslow, a University of California at Davis microbiologist and a leading expert in food safety investigations. "And when you're in that mode, when your objective is to fill orders, you tend to stretch your system—in terms of capacity and throughput, but also in terms of what you can really handle while paying attention to all the details of quality and safety."[17]

Third-generation Japanese American peach grower Mas Masumoto echoes the challenges of reconciling his role as a farmer with his role as a business manager. "Once you start hiring a lot," he recalls his father's advice, "you're not just a farmer anymore." Expenses per acre and yields and aspects of the work that are "easily quantifiable" battle with his awareness that "simple linear formulas do not apply in farming." He minimizes this tension by keeping his farm small enough so that he and his father can do most of the work, except for pruning and harvesting. He speaks of what he calls "synergism," through which the gifts of nature almost magically add up to a whole that is greater than the parts, but he also speaks of war: "I sometimes picture my farm as a battlefield with troops of people struggling with nature in a hundred-year war."[18]

It is a struggle to keep the "culture" in "agriculture." The modern era is replacing "culture" with "business," producing a high-yielding hybrid activity called "agribusiness." This is an activity defined by commodity producers and commodity markets and global trade, by industrialization, mechanization, and what some have

---

17. *The End of Food.*
18. David Mas Masumoto, *Epitaph for a Peach: Four Seasons on My Family Farm* (New York: Harper Collins, 1995).

even called "chemicalization." For multigenerational farm families who have a visceral attachment to and intimate knowledge of a certain piece of land and a certain way of life, or, even, a certain variety of peach, the march toward larger farms and larger markets and larger machines comes at a cost for which no financier can compensate.

Lest we too easily dismiss them as remnants of an agrarian past, let us consider the possibility that such farm families, and the communities of which they are a part, are microorganisms in the soil of the food system. They are the humans who care for the humus. They do not survive, well, the heavy application of the synthetic nutrients of industrial economics. They are as susceptible to arbitrage and futures contracts as an actinomycete to Round Up and Malathion. And without them, not only soil health, but also cultural health, indigenous culture, and local economies—the social and environmental relationships that promote the health of families, communities and bioregions—are all at risk.

We find ourselves, now, between the Scylla of finance and the Charybdis of fertilizer. We are presiding over the unprecedented accumulation of financial capital, on the one hand, and the continuing erosion of social and natural capital, on the other.

We need the kind of reconnoitering that no expert can provide. We need to reconnoiter outside the realms of market share and shareholder, above the top line and below the bottom line, away from soil war and toward soil peace. We need to reconnoiter between our hearts and our minds.

No economist, no soil scientist, no investigative journalist, can show us the way. There is no GPS that will guide the bees back to the hive. There is no price/earnings ratio, no SROI[19] that will show us the way back home. *Do we reckon our whereabouts with economic statistics or earthworm sensibilities? Do we believe more in ownership or usufruct? Chrematistics or oikonomia?*[20] *Lowest price or highest quality? Profitability or fertility? Thomas Malthus or Milton Friedman? How do we make our way between these possibilities?*

My dad used to tell a joke about reconnoitering;

> The Allied offensive reached a river and the commanding general summoned Irving Cohen to his tent.
>
> "Irving, we have crucial mission for you. We need you swim across the river at midnight, reconnoiter the enemy positions, and get back before dawn."
>
> "Yes, sir," Irving replied with his thick Yiddish accent. "I'm de right man for de job."
>
> At midnight, Irving swam off into the darkness. Just

19. SROI stands for "social return on investment," a calculation that seeks to measure both the financial value created and the social impacts of an investment.

20. In *For The Common Good: Redirecting the Economy toward Community, the Environment, and a Sustainable Future* (Boston: Beacon Press, 1989), Herman Daly and John Cobb, Jr. wrote:

> Aristotle made a very important distinction between 'oikonomia' and 'chrematistics.' The former, of course, is the root from which our word 'economics' derives. Chrematistics is a word that these days is found mainly in unabridged dictionaries. It can be defined as the branch of political economy relating to the manipulation of property and wealth so as to maximize short-term monetary exchange value to the owner. Oikonomia, by contrast, is the management of the household so as to increase its use value to

before dawn, he returned, dragging himself up on the river bank, exhausted, his clothes in tatters.

"Irving, what happened? What did you find?" the general asked.

"Vell, on de' right, on the right dey got lots of anti-aircreft, not so much tenks. We go in with a few armored divisions, we can get in dere."

"What about on the left?"

"On de' left, vell, on de' left side, dere dey got lots of tenks, not so much anti-aircreft. We go in with some low flying B-29s, we can get in dere."

"But what about the center? What do they have in the center?"

"In de' center, oh boy, in de' center, forget about it. De' center is absolutely impr-r-regnable," Irving said, shaking his head. "Impr-r-regnable!"

---

all members of the household over the long-term. If we expand the scope of household to include the larger community of the land, of shared values, resources, biomes, institutions, language and history, then we have a good definition of 'economics for community.' . . .

Oikonomia differs from chrematistics in three ways. First, it takes the long-run rather than the short-run view. Second, it considers costs and benefits to the whole community, not just to the parties to the transaction. Third, it focuses on concrete use value and the limited accumulation thereof, rather than on abstract exchange value and its impetus towards unlimited accumulation. . . . For oikonomia, there is such a thing as enough. For chrematistics, more is always better. . . .

It does appear, therefore, that something of a paradigm shift is required in order to admit the Trojan Horse of 'carrying capacity' into the citadel of economic theory. Once that concept is taken seriously a shift in prospect from chrematistics to oikonomia will have begun. Little by little, simple growth in scale of product would no longer be regarded as the combined summum bonum and panacea. That indeed would be a radical shift in outlook.

> "What do you mean?" the general asked. "What could they possibly have in the center that makes it so impregnable?"
>
> "In de' center," Irving said, continuing to shake his head and throwing up his hands in frustration, "In de' center dey got dere a dawwggg, such a dawggg! . . ."

After all the technological wizardry, military-industrial machinations, and global hegemonic workings, history still hinges on the unexpected and the humble, and on a certain immutable recognition of our place in the scheme of things.

The possibility that there is such a thing as *our place in the scheme of things* seems hokey in the age of GPS and private space travel. Yet there is irony in the extent to which we are engineering GPS into every cell phone and every car,[21] while we are losing our sense of where we came from and where we are going in historical and planetary terms.

Our ability to reckon our whereabouts and account for what is going on around us is infiltrated, fragmented, infected, by a virus of numbers, technology, and economics:

- Producer is separate from consumer.
- Investor is separate from citizen.
- Where we live is separate from where we work.
- Communication is confused with thought.
- Diversification is confused with direction.

---

21. A GPS device implanted in a cow's ear is under development, conjuring up the picture of cowboys at control panels, conducting roundups by remote.

- Mobility takes precedence over responsibility.
- We become a nation of commuters and tourists.
- We become a nation of migrant workers and WIMPs.[22]

Belonging nowhere, we are driven by the imperatives of markets from occupation to occupation like bees on a flatbed truck.

When John Maynard Keynes wrote, "Words ought to be a little wild, for they are the assault of thoughts on the unthinking," and James Joyce wrote, deep in the innards of *Ulysses*, "that the language question should take precedence of the economic question," they put their fingers on one of the great wounds of the modern era: We need to discover ways of thinking and speaking that can put economics in its place.

In our devotion to money, market, and machine, we are destroying not only the fertility of the soil, but the fertility of our imaginations.

What is, in the farmer's field, a struggle between economics and ecology becomes, in the investor's mind, a struggle between quantity and quality, portfolios and possibilities, numbers and words.

The good news is that thousands of entrepreneurs and investment advisors and millions of investors have entered the fray over the past few decades, working to cultivate new approaches to capital: social investing or socially responsible investing (SRI), the double-bottom-line (referred to frequently as "doing well while doing

---

22. A WIMP is a white itinerant money person.

good"), the triple-bottom-line (for social, natural, and financial capital, also referred to as people, planet, and profits), mission-related investing by foundations, community-development venture capital, microfinance, social entrepreneurship—these are first steps in the direction of reintroducing whole-earth imagination into the linear formulas of investing and private enterprise.

At the most fundamental level, social investing and social entrepreneurship can be understood as part of an emerging recognition of the limitations, distortions, and seductions of the disciplines and jargon of economic expertise. Integral to this recognition is a more accurate reckoning, a more accurate accounting with respect to the place of the economy in the larger context of history, culture, and natural systems.

The poet laureate of this work is Wendell Berry. Evoking with unparalleled clarity and in unexpected ways the relationship between language and economics, Berry gives new meaning to the ideas of accountability and accounting:

> My concern is for the *accountability* of language—hence, for the accountability of users of language. To deal with this matter I will use a pair of economic concepts: *internal accounting*, which considers costs and benefits in reference only to the interest of the money-making enterprise itself; and *external accounting*, which considers the costs and benefits to the "larger community." By altering the application of these terms a little, any statement may be said to account well or poorly for what is going on inside the speaker, or outside him, or both.
>
> It will be found, I believe, that the accounting will be poor—incomprehensible or unreliable—if it attempts to

> be purely internal or purely external. One of the primary obligations of language is to connect and balance the two kinds of accounting.[23]

Berry does not zero in on the speed of capital, but it is clear that internal accounting is far simpler, far faster than external accounting, which brings with it multiple stakeholders and qualitative distinctions. Connecting and balancing the two kinds of accounting requires patience and care. Severed from external accounting, internal accounting will not only be "poorer," but it will also be "faster."

Neither was the speed of capital central to E. F. Schumacher's argument in *Small Is Beautiful*, focused as he was on the question of scale and related issues of local versus global. He clearly recognized, though, the integral relationship between "size, speed, and violence." Like Berry, Schumacher asserts the precedence of language and qualitative distinctions over economics:

> Economics operates legitimately and usefully within a "given" framework which lies altogether outside the economic calculus. . . . If the economist fails to study meta-economics, or, even worse, if he remains unaware of the fact that there are boundaries to the application of the economic calculus, he is likely to fall into a similar kind of error as that of certain medieval theologians who tried to settle questions of physics by means of biblical quotations. Every science is beneficial within its proper limits,

23. Wendell Berry, *Standing by Words* (San Francisco: North Point Press, 1983).

> but becomes evil and destructive as soon as it transgresses them.
>
> The science of economics is "so prone to usurp the rest"—even more so today than it was 150 years ago, when Edward Copleston pointed to this danger—because it relates to certain very strong drives of human nature, such as envy and greed. All the greater is the duty of its experts, the economists, to understand and clarify its limitations, that is to say, to understand meta-economics.[24]

Meta-economics is not a set of new methodologies. It offers, rather, coordinates that will allow us to chart a new course. To do so, we will need to articulate a new set of values and develop new financial tools that work in concert with them.

We will need to discover new ways of thinking and speaking that go beyond the . . . *hyphenated.*

Meta-economic, socioeconomic, postindustrial, e-commerce, post-consumer: It is no accident that so many of the terms we use to describe our widening worldview are hyphenated or compound. We are splicing new awareness, new insight onto old roots.

Perhaps there is linguistic truth lurking somewhere here: When we get right up against the most important cultural thresholds, when we are being called to the most difficult, most ennobling tasks, we reach for our hyphens and invent new phrases. Traditional terminology fails us. We push toward new meanings, but don't have everything we need, yet, to bring them forth whole cloth.

---

24. *Small Is Beautiful.*

What comes after IRR?[25] *ERR*?[26] What comes after venture capital? *Nurture* capital? What comes after the free market? The *fair* market?

We don't quite know, we are not quite ready to say, and so what remains is to test new phrases or cobble together compound terms that are suggestive, unable to let go of the old and fully embrace the new.

The curious ambiguities of compound words and hyphenation are not confined to matters economic, of course. Consider, for example, the case of *non-fiction*.

We do not call the literary genre to which this term refers "truth" or "truth telling." We say only that it is "not" fiction. There is something fascinating about this linguistic choice. We give that which "isn't" a name, *fiction*, and then describe that which "is" only as a kind of double negative, *non-fiction*, as in "isn't-isn't," as if recognizing, implicitly, that there is an element of fiction in any depiction. It's as if, deep down, although we don't want to openly admit it, we know that we do not know. No fact can be completely accurate or accurate unto itself. There is no such thing as "value-free" science or "value-free" economics, despite the comfort offered by numerical reductionism. We can never extricate ourselves completely from the fiction-influenced lens or observer bias. This is why the uncertainty principle is so resonant as an explanation of ambiguity that exists within science, and, in broader ways, along the boundaries of science and nonscience.

---

25. IRR stands for "internal rate of return," the time-sensitive metric used by professional investors to calculate financial returns.

26. ERR stands for "external rate of return," a metric that does not yet exist, except as an imaginative reference to the long-term social and environmental impacts on communities and bioregions that are typically "externalized" by companies as they maximize growth and maximize returns to their shareholders. As in the meta-economic dictum, "To IRR is human, to ERR divine."

Which raises the question: Why don't we call science "non-religion" or "non-uncertainty studies"? Or economics "non-meta-economics"?

In "The Role of Economics," the chapter in *Small Is Beautiful* in which Schumacher introduces the concept of meta-economics, he also uses the following hyphenated terms: *non-economic*, *non-responsibility*, *profit-making*, *well-being*, *God-given*, *soul-destroying*, *home-produced*, *non-growth*, *self-deception*, *economist-turned-econometrician*, and *market-oriented*. Not accounting for multiple uses of the same term, and ignoring the fact that a double-hyphenated term could be counted twice, that's a very high HT/P (hyphenated term per page) ratio of .92.

It is as crude to pluck a passage from this marvelously coherent little chapter as it is to jam a hyphen between the words *profit* and *making*, but I'll pluck one, anyway:

> The market therefore represents only the surface of society and its significance relates to the momentary situation as it exists there and then. There is no probing into the depths of things, into the natural or social facts that lie behind them. In a sense, the market is the institutionalization of individualism and non-responsibility. Neither buyer nor seller is responsible for anything but himself. It would be "uneconomic" for a wealthy seller to reduce his prices to poor customers merely because they are in need, or for a wealthy buyer to pay an extra price merely because the supplier is poor.[27]

27. *Small Is Beautiful.*

Now, this quote is apt to take the brunt of free-market ideological retort. To which I would offer the following preemptive attack: The point is not to be ideologically *against* markets any more than to be ideologically incapable of recognizing that markets are not a panacea. What we are seeking is a "connected" and "balanced" frame of reference that is greater than, but includes and respects, the power of markets. Similarly, meta-economics is not *against* economics; it is against economics that lacks social and environmental grounding. (It is often pointed out by students of Adam Smith that in addition to *An Inquiry into the Nature and Causes of the Wealth of Nations*, Smith also wrote *The Theory of Moral Sentiments*, which explored the web of sympathies, virtues, and duties that provide an ethical context for the workings of the Invisible Hand of the marketplace.)[28]

We might call Schumacher an economist-turned-philosopher or a fiduciary-turned-activist. Or can we just call him a sage and go . . . hyphen-naked?

All of this hyphen play is heading somewhere.

We are heading toward the mother of all once-hyphenated destinations: nonviolence.[29]

---

28. *The Theory of Moral Sentiments* opens: "How selfish soever man may be supposed, there are evidently some principles in his nature, which interest him in the fortunes of others, and render their happiness necessary to him, though he derives nothing from it, except the pleasure of seeing it." Smith saw such affections as co-equal to the kind of enlightened self-interest for which *The Wealth of Nations* has enshrined him. He wrote: "To feel much for others and little for ourselves, . . . to restrain our selfish, and to indulge our benevolent, affections constitutes the perfect of human nature, and can alone produce among mankind that harmony of sentiments and passions in which consists their whole grace and propriety." (*Moral Sentiments*, part 1, sec. 1, chap. 5)

29. After years of common usage, once-hyphenated terms often lose their hyphens.

Although Berry's advocacy revolves around agrarianism and Schumacher's around appropriate scale and appropriate technology, both men are part of the broader historical movement toward the possibility that one day, nonviolence might trump violence as an organizing principle for the affairs of man.

In an era dominated by all things financial and quantitative, social investing constitutes the beginning of a reawakening, a reconnoitering. We do not have a term that fully describes our next destination—"Restorative Economics" seems as good a placeholder as any—yet we recognize its core principles: *carrying capacity, cultural and biological diversity, sense of place, care of the commons, and nonviolence.*

Of these, the last, nonviolence, is the most fundamental.

Organic farming is about growing food without doing harm to natural systems and producing food that does no harm to its eaters. Renewables and clean tech are most obviously about keeping carbon out of the atmosphere, but they are also about generating energy in ways that do no harm to communities and bioregions. Independent media is about packaging and delivering information in ways that do not destroy cultural diversity. Alternative and integral medicine are about healing that does not rely on pharmaceuticals and surgery, but rather on harm-free modalities of prevention and wellness.

All of these are steering away from industrial-scale, mega-consumption, global-brand-dominated, technology-first, growth-maximizing economics and toward an economy that operates on a scale that is more humane and more comprehensible, at a speed that is less dizzying, in a manner that is less extractive and less hell-bent for growth—in other words, less violent. Organics, renewables, independent media, and integral medicine are all responses to the inherent violence of the modern economy, which has proven a phenomenally effective machine for growing capital, but a predictably blunt instrument when it comes to healing, preservation, and restoration.

To see the underlying violence of the modern economy in relief, we must look beyond Buffett and Gates and Soros and Malkiel and Graham,[30] to the likes of Berry and Schumacher and Gandhi and Tolstoy and Thoreau. We must look beyond the wizardry of financiers to the wisdom of meta-economists:

> As with our colleges, so with a hundred "modern improvements." The devil goes on exacting compound interest to the last for his early share and numerous succeeding investments in them. Our inventions are wont to be pretty toys, which distract our attention from serious things. They are but improved means to an unimproved end.[31]

> People who own great estates or fortunes, or who receive great revenues drawn from the class who are in want even of necessities, the working class, as well as all those who like merchants, doctors, artists, clerks, learned professors, coachmen, cooks, writers, valets, and barristers, make their living about these rich people, like to believe that the privileges they enjoy are not the result of force, but of absolutely free and just interchange of services, and that their advantages, far from being gained by such punishments and murders as took place in Orel and several parts of Russia this year, and

30. Burton Malkiel's *A Random Walk Down Wall Street* and Benjamin Graham's *The Intelligent Investor* are widely considered bibles of modern investing.
31. Henry David Thoreau, *Walden*.

are always taking place all over Europe and America, have no kind of connection with these acts of violence. They like to believe that their privileges exist apart and are the result of free contract among people; and the violent cruelties perpetrated on the people also exist apart and are the result of some general judicial, political, or economical laws.[32]

It is difficult but not impossible to conduct strictly honest business.[33]

Greater even than the mystery of natural growth is the mystery of the natural cessation of growth. There is measure in all natural things—in their size, speed, or violence. As a result, the system of nature, of which man is a part, tends to be self-balancing, self-adjusting, self-cleansing. Not so with technology, or perhaps I should say: not so with man dominated by technology and specialization. Technology recognizes no self-limiting principle—in terms, for instance, of size, speed, or violence. It therefore does not possess the virtues of being self-balancing, self-adjusting, and self-cleansing. In the subtle system of nature, technology, and in particular the super-technology of the modern world, acts like a foreign body, and there are now numerous signs of rejection.[34]

We should recognize that while we have extravagantly subsidized the means of war, we have almost totally

32. Leo Tolstoy, *The Kingdom of God Is Within You.*
33. Mahatma Gandhi, *Non-Violence in Peace and War.*
34. *Small Is Beautiful.*

> neglected the ways of peaceableness. We have, for example, several national military academies, but not one peace academy. We have ignored the teachings and the examples of Christ, Gandhi, Martin Luther King, and other peaceable leaders. And here we have an inescapable duty to notice also that war is profitable, whereas the means of peaceableness, being cheap or free, make no money.[35]

Can nonviolence coexist with profitability? Can "doing no harm" become integral to industry? Can the modern investor, inured as he is to externalities and all things uneconomic, and awash in the wealth created by a once-in-the-history-of-the-planet confluence of population growth and technological innovation, recognize and redress the destructiveness of the modern economy?

If the Gandhian and the Tolstoyan and the Thoreauvian seem too at odds with modern notions of entrepreneurship and fiduciary responsibility to provide useful reconnaissance, consider that thousands of small companies and investors are, in fact, part of an emerging "do less harm" industry.

As just one data point, consider Investors' Circle, a U.S. network of angel investors, family offices, foundations, and social-purpose funds investing in early-stage companies that create commercial solutions to social and environmental problems. Since 1992, Investors' Circle members, under the banner of "patient capital for a sustainable future," have invested more than $130 million in more than 200 early-stage companies and social-purpose

35. Wendell Berry, "Thoughts In the Presence of Fear," web exclusive to the Autumn 2001 issue of *Orion Magazine.* www.orionmagazine.org/index.php/articles/article/214

venture funds including: Verdant Power (tidal electricity generation from submersible turbines), Farmers Diner (family fare sourced organically and locally), SPUD (organic food home delivery service), IceStone (high-end architectural surfaces of recycled glass and cement, manufactured in an empowerment zone), Organic to Go (organic convenience food and corporate catering), Energia Global (low impact, high-head hyrdroelectricity in Latin America), Sonic Innovations (breakthrough hearing aid technology), ZipCar (European-style urban car rental by the hour), Teach First (professional development educational software), Wild Planet Toys (educational, nonviolent toys), MicroBanx (microfinance services), United Villages (internet connectivity for rural villages in developing countries), Voxiva (disease outbreak early-alert hardware and software), TerraCycle (liquid worm-poop fertilizer, mass marketed), Coast of Maine Organics (composted soil products from aquaculture and agricultural by-products), Sun Edison (solar energy services to government, utility, and commercial customers), CitySoft (software developer creating jobs in the inner city), and Lumenergi (low-energy light ballasts).

A "patient capital" marketplace is emerging to better serve such companies, since most are not easy candidates for the same dollars that are seeking the next Google.

Patient capital does not exist yet as an organized or disciplined asset class; it is the gestalt that emerges as socially responsible investing matures and as the wave of triple-bottom-line entrepreneurs and investors builds. The screened mutual funds and shareholder advocacy of socially responsible investing are early indicators of deeper upwellings.

Applied to the food sector, patient capital becomes slow money—whose name carries with it more than a doff of the cap to

Slow Food, the international NGO that promotes biodiversity, artisan food traditions, heirloom varieties, and connections between small farmers and consumers. Slow money can be thought of as a subsector, a sub-asset class of patient capital, focused with appropriate patience on the health of soil and bioregion.

Slow money is patient capital on the opposite of steroids.

Asserting a connection between such luminaries as Thoreau and Tolstoy and Gandhi and a bunch of early-stage companies in the first decade of the twenty-first century runs the risk of seeming naïve, as well-intentioned but ineffectual as an old hippy at an annual shareholders meeting of Intel.

Yet social investing can best be understood, with its historical roots in Quakerism and anti-apartheid divestitures, as an expression of a centuries-old ethos of nonviolence in the context of modern fiduciary capitalism. Of necessity, this expression manifests itself in partial adaptations, pragmatic mutations, and imperfect applications. Lots and lots of half steps. After all, who can ignore how daunting it is to look at the Fortune 500 or the Russell 5000 and think: What would I invest in if I *really* wanted to do no harm?

Our "inescapable duty," to use Wendell Berry's words, is to avoid acting like deer in the headlights, and to move forward, undeterred by ambiguity. Our success in moving beyond half steps, in finally defining a new destination and taking the first full steps in its direction, depends on acknowledging, without scapegoats and without undue recrimination, the violence of the modern economy.

By prioritizing markets over households, community, place, and land, the modern economy does violence to the relationships that underpin health and that give life-sustaining meaning—family relationships, community relationships, relationships between consumers and producers and between investors and the enterprises in which they invest, relationships between companies and the places in which they do business, relationships to the land and in the soil. Such relationships are attenuated, or, in the extreme, deracinated, by the modern, global economy.

The extent to which the modern economy depends on broken relationships was revealed by a story told a few years ago during a small retreat of business leaders. Going around the table to introduce one another, a young entrepreneur of Middle Eastern descent told a tale of broken relationships:

> My most recent software company had its offices in the World Trade Center. On 9/11, we were pretty much wiped out, most of our records gone. When we started trying to put some of the pieces back together, I made the rounds to my directors. These were many of the leading investment bankers on Wall Street, individuals with whom and for whom I had made many, many millions of dollars in my previous ventures.
>
> One of them said to me, "Why don't you have Osama fund your re-start?"
>
> At that moment, I realized that I had no relationships with these people. I realized that there had been nothing but commercial ties between us. The money connections were not real relationships.

As he spoke, it became clear that while many of us had been aware for some time of the manner in which the modern economy depends on and produces broken ecological relationships, we had not been fully cognizant of the corollary damage done to social relationships. Although we understood the concept of ecological footprint, we did not fully understand the footprint of broken social relationships.

It is no accident that an economy based on broken relationships would find it easy to support, and to depend on, the building of nuclear weapons, the waging of wars in distant lands, the selling of cigarettes, the flying of trillions of air miles, the commodification of leisure, urban and suburban sprawl, gated communities and *favelas*, toxins in the food and water, and kids who watch an average of four hours per day of TV, paying more attention to it and to instant messaging than to people in the room. Much of this violence is overt. Much of it is implicit, indirect, made palatable by the fundamentals of consumerism and made invisible by veils of intermediation.

As it has been practiced and understood, socially responsible investing can do little to address root issues of consumerism and intermediation. SRI is confined largely to damage control: an exercise in improving corporate governance and minimizing damaging "externalities," while not affecting core elements of a company's activities, culture, or mission. At its worst, SRI seems an exercise in fueling a bulldozer with biodiesel: We are greening our fuel, but are we preventing subdivision of the farm?

Despite successes of shareholder-advocacy campaigns and occasional spikes in public awareness, there is little in the nature of fundamental systemic change that can be accomplished via broadly diversified portfolios of mature public companies, managed by SRI brokers and advisors who compete with the financial returns and performance benchmarks defined by the very same extractive or destructive economic activities that are the objects of SRI reform.

To look at some of the growth statistics provided by the Social Investment Forum, you would conclude that SRI is making substantial inroads into the capital markets. From 1995 to 2005, assets under management using one or more of the three core socially responsible investment strategies—screening, shareholder advocacy, and community investing—rose from $639 billion to $2.29 trillion. During that same period, the number of socially screened mutual funds rose from 55, with assets of $12 billion, to 201, with assets of $179 billion.

Despite the dramatic increase in the number of socially screened mutual funds, however, the overall increase in SRI assets is far less impressive when viewed against the increases that occurred in total investment assets under professional management in the United States from 1995 to 2005—from $7 trillion to $24.4 trillion. Even if one considers SRI's percentage of total assets, roughly 10 percent, to be impressive, and even if one takes at face value a recent study by Mercer Consulting indicating that three-quarters of all money managers think that social investing will pervade the financial sector within a decade, the imperfections inherent in SRI cannot be ignored.

Critics of SRI point out that its pursuit of competitive returns leads inevitably to watered-down screens, resulting in the fact that some 90 percent of the Fortune 500 companies make it into an SRI portfolio. For instance, a fast-food company or an oil company or a mining company may be deemed "best of class" for some of their corporate governance practices, but this does not address the basic problem that Paul Hawken identifies: "If you are going the wrong way, it doesn't matter how you get there."

The core issue that the SRI industry is not confronting is the macro problem of economic growth, which manifests itself at the micro level of individual portfolios and individual investments as the problem of competitive returns. Professional managers are measured on their financial performance, and it is no different for those who incorporate social and environmental criteria. The result: elaborate strategic machinations and metrics designed to demonstrate that you can "do well while doing good," which in other parlance might be called "having your cake and eating it too."

"The industry has hooked people on the idea," Hawken wrote in a controversial 2004 critique, "that SRI funds should do as well as or better than other mutual funds, and then they have to demonstrate it, which leads to portfolio creep—the dumbing down of criteria and the blurring of distinctions between what is or is not a socially responsible company."

To fully consider the social and environmental responsibility of a company, qualitative distinctions and exogenous factors must be considered with respect to its business. Is a company promoting conspicuous consumption? Is it furthering a global brand at the expense of local enterprise? Is it directly or indirectly exacerbating rural-urban migration? Is it benefiting from the utilization of natural resources at an unsustainable rate? Is it reducing cultural and biological diversity?

We must ask such questions humbly, never forgetting that, in Aristotle's words, the perfect is the enemy of the good. We must be critical but not too critical of the environmental and governance-related failings of mature corporations. We must be critical but not too critical of the environmental and governance-related failings of start-up companies that explicitly embrace the triple-bottom-line.

At the Jessie Smith Noyes Foundation, one of the pioneers of mission-related investing during the 1990s, finance committee banter often circled around the "Viederman Three": the fictional three companies that were pure enough to pass muster with the foundation's president, Stephen Viederman. Such banter expressed humility at the difficulties of clearly posing and answering questions about socially responsible business practices.

In mature public companies and venture capital-backed companies, with the mandate to maximize shareholder returns built in like a kind of permanent pedal to the metal, response to questions of social and environmental responsibility are highly constrained. Philanthropy is similarly constrained in its approach to such questions, as the culture and imperatives of grant making define agendas and priorities. What does that leave? Up until now, that has left only SRI mutual funds and NGOs, fighting rear-guard actions against particular companies or particular corporate practices—an advocacy posture that entails pushing a very big stakeholder rock up a very steep fiduciary hill.

Such advocacy inevitably depends more on sticks—imposing fixes on companies through negative feedback—than on carrots, since NGO-led advocacy is inherently adversarial and investor-led advocacy in almost no cases represents more than a tiny percentage of a mature corporation's capitalization.

But it all comes down to the carrots.

If you are not a shareholder (and in a very few cases even if you are), it is easy to rant against the excesses, omissions, malfeasances, lack of full accountability, and unenlightened self-interest of global corporations.

Thank Gaia, it is also easy to plant carrots.

As figurative carrots, take the thousands of companies that are started up each year by entrepreneurs who see their companies as agents of cultural and environmental restoration. Fair trade companies. Inner-city child care companies. Solar and wind companies. Educational ventures. Publishing ventures. Companies developing medical products and services for niche markets and developing countries. Companies whose products and services promote health and wellness. And, of course, organic food companies. They are "carrots" because they offer a forward-looking, non-adversarial, even celebratory complement to advocacy "sticks." Providing start-up capital and growth capital to such enterprises is like planting carrots.

But we cannot talk about carrots as metaphor without also talking about non-fiction carrots. Take those, for instance, of one Eliot Coleman, who, with his wife, Barbara Damrosch, and a steady crop of interns who come from near and far, runs Four Season Farm in Harborside, Maine.

These carrots, the smell and texture of the soil in which they are so densely planted, the almost meticulously tended beds they are grown in, their hue and translucence, their sweetness—all conspire to arrest the unsuspecting consumer, in much the same way as Brunello sipped in Montepulciano (at roughly the same latitude, 44 degrees north or so, which is very surprising to most folks, given how different the climates are in mid-coastal Maine and Tuscany). They give such pleasure and awaken the senses, suggesting questions that have everything and nothing whatsoever to do with socially responsible investing: *When did we forget our connection to the land? How can this taste so damned good? What* ***are*** *all these tastes, anyway? Why doesn't all food taste like this? How much longer would I live if I ate like this* *every day? Who cares about longevity—that's another numbers thing! This is all about quality?! How does he tend this place so artfully? When I pull one of these carrots, why does the earth release it so easily? Why does the earth smell and feel so good? Where was it that I was in such a rush to go yesterday? When did we lose our way? Can I take you home with me? How about a hug? Why can't I live* ***here****?*

Of course, everyone can't live on an organic farm. And everyone isn't Eliot Coleman any more than everyone is Joni Mitchell.

But that doesn't mean we all have to keep doing the same thing over and over again, hoping for a different outcome. We don't have to keep sending our money into distant, invisible portfolios, while wondering why Main Street is dying, our food is irradiated, and geneticists in China are breeding square apples.[36] We can find ways to build the soil of local food systems.

This is the beauty of Slow Food. It gives us a way to engage that is proactive, and even celebratory. It taps into and follows a beautiful energy that shifts, quickly and fundamentally, away from anti-McDonald's and anti-globalization and toward pro-heirloom variety and pro-*terroir*[37] and pro-artisan and pro-small farmer, pro-biodiversity and pro-localization and pro-community.

This can also be the beauty of slow money.

There is beauty in the small, independent, local-first food enterprises that are candidates for the kind of support that slow money could provide.

---

36. That's right. Entrepreneurship takes all forms: in this case, square apples, designed to optimize packing and shipping.

37. *Terroir* is a term referring to specific food qualities and taste that derive from a particular region or locale.

Consider, for illustrative purposes, the following enterprises as exemplars of the kind of "SFEs" (small food enterprises) that could be candidates for slow money. These examples come from northern Vermont, where we have recently conducted field research to inform the modeling of a Slow Money portfolio:

- Butterworks Farm. Thirty-year-old regional organic yogurt brand.
- High Mowing Seeds. Twelve-year-old certified organic seed company.
- Jasper Hill Farms. Early-stage organic cheese maker and cheese cave.
- Pete's Greens. The largest organic CSA in Vermont, and growing.
- Vermont Smoke & Cure. Regional pork processor and wholesaler.

These SFEs play a vital role in their local food system and prioritize this role in their mission. Each has an established customer base and reputation for product excellence. Each is desirous of what we can refer to, without tongue completely in cheek, as "organic" growth: They wish to expand their impact, but they are opposed to growth forced by outside capital and they abjure an "exit strategy" that would put control of their business in the hands of absentee shareholders. None is much bigger than $1 million in sales. They are too big for micro-finance and far too small for venture capital. They are not easy candidates for equity capital. And because they are for-profit businesses, they are not easy candidates for philanthropic support.

In a different category in terms of type and stage of enterprise is Carbon Farmers of America, which, while located in the region, has a national business strategy, seeking to commercialize organic soil-building methods and link soil fertility to carbon-trading payments for farmers. The company is very early stage, has no established track record, but has the potential for significant scalability.[38]

The northern Vermont region is also home to many restaurants that are committed to sourcing food as locally and organically as possible. These include Claire's in Hardwick, The Bee's Knees in Morrisville, The Hen of the Table in Waterbury, Farmers Diner in Quechee, American Flatbread in Waitsfield, and Smokejack in Burlington. Owing to scale, profitability, and other risk factors, restaurants do not even show up on professional investors' or philanthropists' maps as an organized asset class, although from the standpoint of their impact on the health of local food communities, they often play important roles.

In identifying the types of enterprises that would be consistent with slow money, we must also look outside any particular region, since distribution, marketing, and processing companies working across regions play important roles in building infrastructure that supports the overall development of sustainable food systems. We might consider such companies part of the broader class of SFEs that we are in the process of targeting:

---

38. The potential of soil rich in organic matter to sequester carbon deserves much, much more than a footnote. A 27-year-long study by the Rodale Institute found that organic farm soil can sequester 1,000 kilograms of atmospheric carbon per hectare per year. In Australia, proponents of the Soil Carbon Manifesto believe that soil carbon sequestration has the potential to substantially counteract, on a global basis, greenhouse gas emissions.

- Organic Valley. This cooperative with 1,300 U.S. dairy farm members is a national leader in the production of organic milk and cheese.
- Equal Exchange. For the past twenty years, this hybrid cooperative has been a leader in fair-trade coffee.
- Guayaki. This eight-year-old company pioneered the U.S. market for yerba maté, a drink made from herbs sustainably harvested in the Paraguayan rainforest.
- Theo Chocolates. This early-stage company is the nation's first roaster of certified fair-trade organic cocoa beans.
- Marrone Organic Innovations. This early-stage company develops environmentally responsible natural products for weed, pest, and plant-disease management.

Many thousands of SFEs are directly or indirectly synergistic with the work of building healthy food systems in any given region. Approximately 1,000 organic food companies in the United States have sales of between $2 million and $10 million. There are close to 2,000 CSAs nationally, up from only a handful twenty years ago.[39](The largest CSA in the United States has 2,000 members; the largest CSA in Denmark has 50,000 members and revenues of approximately $50 million.) Hundreds of companies are developing inputs and supplies for organic farming, including soil amendments, fertilizers, and ecologically benign pest-management products. Farmers markets are up nationwide from

39. CSA, which stands for "community supported agriculture," describes a farm in which consumers buy shares of the year's produce in advance. In most U.S. CSAs, produce is picked up on designated days of the week at the farm, and consumers get their share—say, for example, 1/100th at a CSA that has 100 members—of whatever the farm has harvested at that time.

1,755 in 1994 to 4,385 in 2006. Regional natural and organic food markets and co-ops continue to find vibrant niches: City Market in Burlington, Vermont; New Seasons in Portland, Oregon; Cid's in Taos, New Mexico; Harvest Coop in Cambridge and Jamaica Plain, Massachusetts; and Rainbow in San Francisco. Sales of organic food in the United States have been growing at approximately 20 percent per annum for a few decades, now reaching some $20 billion in 2007, still less than 3 percent of all food sales.

Despite the continuing overall decline in the number of U.S. farms, and the loss of at least 1 million acres of farmland to development each year, a 2002 census report indicated growth in the 10- to 49-acre-farm category: Between 1997 and 2002, the number of such farms increased by 33,000.[40] Chef Dan Barber calls the reemergence of small, diversified farms "the greatest upheaval since the Green Revolution," driven not by "a nostalgic bid to revert to the agrarian ways of our ancestors" but by the imperatives of "looking toward the future, leapfrogging past the age of heavy machinery and pollution."[41]

Although the SFE is not recognized by economists or policy makers as a discrete business category, in terms of number of employees or revenues, MSMEs—micro, small, and medium-sized enterprises—are widely recognized as critical to overall economic health. Definitions vary. India's Ministry of Micro, Small, and

---

40. Elizabeth Henderson, *Sharing the Harvest: A Citizen's Guide to Community Supported Agriculture* (White River Junction, VT: Chelsea Green Publishing, 2007). In revenue terms, the USDA calls farms "small" that have revenues of less than $250,000, making only 8 percent of all U.S. farms, by such reckoning, "not small." Sixty-two percent of U.S. farms have sales of less than $20,000.

41. Dan Barber, "Change We Can Stomach," *New York Times*, May 11, 2008. Barber reports that a 4-acre farm nets an average of $1,400 per acre, while a 1,364-acre farm nets an average of $39 per acre.

Medium Enterprises designates specific product categories that are "reserved" for small-scale companies; it is estimated that 12.5 million MSMEs produce 50 percent of the country's industrial output. How are these MSMEs defined? India uses revenues: Micro does not exceed 25 lakh rupees ($53,000); small does not exceed five crore rupees ($1.1 million); medium does not exceed ten crore rupees ($2.1 million). In the United Kingdom, the Department of Business, Enterprise, and Regulatory Reform defines a micro firm as one having up to nine employees and revenues of up to £2 million ($3.5 million), a small firm as up to 50 employees and revenues of up to £10 million ($17.8 million), and a medium-sized firm as up to 250 employees and revenues of up to £50 million ($89.2 million). In Canada, an SME is considered a business with less than 500 employees and CDN$50 million (US$46.6 million) in revenues.[42] Similarly, the U.S. Small Business Administration defines small businesses as those with less than 500 employees, noting:

> To many, the small business owner is synonymous with small town America and an alternative to large multinationals.
>
> The importance of small business is not just an American phenomenon. The Bologna Charter on SME Policies adopted on June 15, 2000, by more than 45 countries recognized . . . "the increasing importance of small and medium-sized enterprises in economic growth, job creation, regional and local, and social cohesion."[43]

42. All conversions to U.S. dollars calculated on September 15, 2008.
43. SBA, "The Governments' Role in Aiding Small Business Federal Subcontracting Programs in the U.S." September 2006.

U.S. Census data shows that 60 to 80 percent of net new jobs are created by small businesses. Yet the provision of capital is skewed to high-tech start-ups, on one end of the continuum, and microenterprise in developing countries, on the other. Less than 1,000 venture capital firms steer roughly $20 billion into 3,000 or so companies per year in the United States; more than 200,000 U.S. angel investors (high net-worth individuals investing their own money) steer roughly the same amount into another 50,000 high-tech companies. At the other end of the continuum, financially and geographically, is the micro-finance industry, which provides loans of between $100 and $1,000 or so to micro-entrepreneurs among the world's poor in developing countries. Micro-finance intermediaries are growing at what Forbes calls a "fever" pace, with 40 new funds started since 2005 and some $17 billion in loans outstanding worldwide.

Meanwhile, back at home in the country that has gotten used to being called "the world's breadbasket," the category "small food enterprise" does not exist. Capital markets seem ready to leave SFEs unrecognized and languishing, neither fish nor fowl in the territory between investing and philanthropy, between developed global economy and developing local economies.

In *The Soul of Capitalism*, William Greider heralds the importance of "thousands of small and independent enterprises pursuing the new ideas and nature-friendly products" that will shape the future of the economy, and points to scores of farmer-owned enterprises that are developing new household cleaners, waxes, water-based resin paints and coatings, lubricants, and other products from soybean, canola, and other oils. Paul Hawken echoes Greider's sentiments: "Ecological restoration can probably be carried out more naturally and surely by smaller enterprises, than by larger, unwieldly corporations. The diversity of the small business sector must be encouraged. . . . [We must] liberate the imagination, courage and commitment that resides within small companies."[44]

Pointing out that the average large company is 16,500 times the size of the average small company, and that the 1,000 largest companies account for 60 percent of America's GNP, with the balance accounted for by 11 million small business, Hawken concludes: "We humans have yet to create anything that is as complex and well-designed as the interactions of the microorganisms in a cubic foot of rich soil."[45]

Perhaps the economic activity that comes closest to the spirit that Hawken is invoking is a CSA. Compared to the rest of the industrial food system, CSAs are simple, small, diversified, direct, and local, balancing financial, social, and natural capital in ways that are beyond, or below, the capacity of most commercial enterprises.

CSAs link farmers with a few dozen, a few hundred, or, in a very few cases, a thousand or more local consumers. In commercial terms, they are tiny. The Robin Van En Center for CSA Resources lists 1,208 CSAs in the United States, with other estimates indicating 1,500 to 2,000, with more than 100,000 total shareholders—with *share* meaning, literally, *share of the harvest*. If we use $500 to $750 as a rough estimate of average share cost, this means that total CSA gross revenues in the United States come to roughly $50

44. Paul Hawken, *The Ecology of Commerce* (New York: HarperCollins, 1993).

45. Ibid. Since 1993, business concentration has continued. In 2007, the largest 1,000 companies accounted for 86 percent of the GNP, and roughly six million smaller businesses accounted for the rest.

to $75 million. Median CSA farm size is fifteen acres, with seven acres of cropland.

Is it socialism? Is it capitalism? CSAs defy such choices.

"Microbial life in the soil is our bank. If we extract life faster than we reinvest, the soil becomes finite," says Paul Muller, one of the owners of Full Belly Farm, in Guinda, California. Full Belly's CSA, more than fifteen years old, delivers its shares, which cost $17 per week, to approximately 1,500 shareholders, mostly in the Bay Area. "Our customers are also our bank." He continues:

> They are our storehouse of financial capital. What is really important about CSAs is that these relationships allow us to leave behind the export model, which for years was all there was in farming. You grew food, but you went to the market to buy your groceries. In fact, one of our neighbors has been raising cattle for 50 years, and until last year he had never eaten any of his own beef. In the old export model, you never knew the buyer. You didn't realize the full value of your produce, so you couldn't reinvest. With the CSA model, we are in a position to reinvest, to keep money moving locally, and this has a revitalizing effect on rural communities that desperately need it. More than that, the CSA creates direct feedback loops between consumers and the growing of their food. The value of these information-rich relationships is hard to overstate.

So, we begin a crude taxonomy of the restorative economy and the local food systems that are critical to its health: small farms, small food enterprises, a culture that values relationships and qualitative distinctions as much as it values transactions and metrics, soil that is valued for its organic matter and biodiversity, food that is valued for its freshness and absence of toxic residues, communities that value making a living over making a killing, investors who value a carrot in the hand as much as two futures contracts in the bush.

If family farmers are the microorganisms in the soil of the restorative economy, then local entrepreneurs are its earthworms. We don't quite know what to make of either of them in purely financial terms, just as we don't yet understand much about what Mas Masumoto called the *synergisms* of the farm. But we know how important they are to the cultural and ecological web, just as a farmer who knows his land can tell so much—though the science be yet in its infancy—from the feel, the smell, and the taste of his soil.

It just may be that all the financial metrics and taxonomic sophistication in the world will not compete for meaning, and for direction, with the feel, the smell, and the taste of the soil. It is not easy to say such a thing with a straight face in this day of professional disciplines and computational firepower. We find ourselves in the position of having to fight our way back, through veils of urbanization and industrialization and securitization and institutionalization, to the most basic of insights, the most basic of affirmations.

Setting out to design financial intermediation from the ground up, rather than from markets down, our way is marked by a number of hypotheses.

- Hypothesis One: *Thinking in terms of bioregions and communities is necessary, but not sufficient: We must zero all the way in, down to the land itself, to our final destination and the source of our enduring wealth and health—soil fertility*. Therefore, SFEs as an asset class cannot be the whole of our investment focus. A regional approach to food enterprises will remain superficial, and, in the long run, ineffectual if it lacks integral investing in farmland itself, keeping land in farming and facilitating its transition to organic practices; and in timberland,

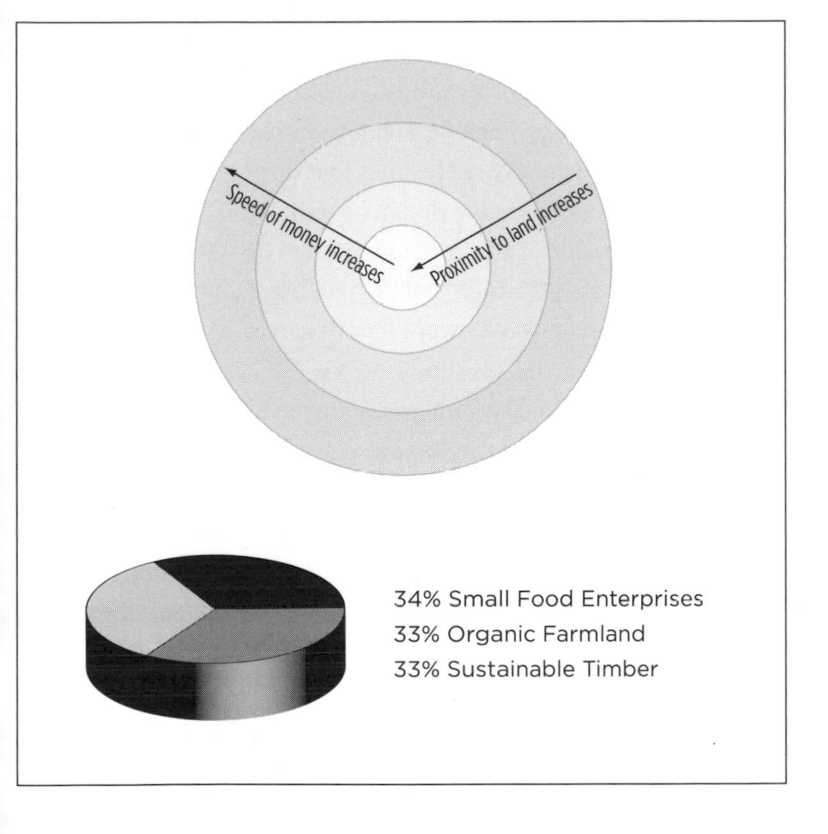

supporting the use of sustainable forest-management practices that promote water percolation and aquifer recharge. Presented visually, this affirmation looks something like the charts on page 61.

- Hypothesis Two: *Although SFEs demonstrate certain characteristics as a class, each individual SFE is too small and too risky as an individual investment.* The risk mitigation of a portfolio works powerfully in high-tech venture capital, where any one investment has very low chances of success but has the potential to be a "home run" that makes everyone associated with it fabulously wealthy. This makes a typical venture capital portfolio a bit like a dozen or two forced death marches or failed missile launches: Each deal shoots for the moon but few come anywhere close. The ones that don't are triaged, sold for pennies on the dollar, or written off. The financial value of the portfolio is often driven by one winner.[46] This is a high-stakes, high-risk game, with all of the adrenaline-rich trappings of gambling.

  Why on earth—or, should we say, "of" earth—would anyone want to take similar risk just to help a CSA grow from $100,000 to $1 million in revenues or a cheese maker grow from $1 million to $5 million in sales?

- Hypothesis Three: *The synergisms embedded in the local and regional relationships created by SFEs may reduce market risk.* "There is no question in my mind that what is going on here, in this region," says High Mowing Seeds

46. A veteran venture capital investor once revealed during a workshop that in 25 years of investing, his overall returns were highly dependent, across several venture funds that he had raised and managed, on the success of one investment.

CEO Tom Stearns, "is creating a kind of business ecosystem that adds value to all of these small enterprises. I see it every day, and even if it is difficult to quantify, or quantify in a way that means something to folks whose only financial realities have seven or eight or nine zeros in them, I know that the kind of value we are creating here is going to be crucial for the health of our children and their children." He adds, "We are not isolationists or anticorporate activists. Not at all. I'm looking to make money from my business. I do business with folks across the United States. But there is something about small-scale food-related businesses that creates a real sense of connection. I have a different kind of relationship with my suppliers and customers, and with the other small business folks in this region. We are playing a kind of stewardship role that is palpable. I can feel the difference we are making in this place and that means a lot to me." High Mowing Seeds, one of the few 100 percent certified organic seed companies in the United States, has grown to $1.2 million in vegetable-seed sales to 3,000 customers nationally over a decade; the company recently raised $800,000 to support its growth to $3 million in sales over the next five years. Is it possible that a portfolio of such companies will offer investors demonstrably lower risks over time? (It took Stonyfield Farm and Odwalla approximately ten years to reach $10 million in sales. It took Butterworks Farm approximately 25 years to reach $1 million in sales.)

- Hypothesis Four: *Although SFEs are for-profit enterprises, our definition of investing may need to be*

> *expanded to describe and support a new set of relationships*. This kind of investing will be as different from buy-low/sell-high investing as a CSA or farmers market is from a superstore. We will need to explore new forms of intermediation that reflect these qualitative differences. Perhaps we need a new tier of philanthropies, set up with the explicit purposes of investing all of their assets in local food systems and local economies. Perhaps every time someone checks out of the supermarket or co-op, they can be given the opportunity to contribute a dollar to the Slow Money Common Fund. Perhaps we need to invent a new kind of municipal bond, dedicated to investing in local food systems.

These hypotheses test aspects of an upside-down investment approach designed to prioritize places over markets. In their formulation and testing, we recognize the importance of investing in ways that affirm the primacy of social and environmental relationships on which lasting financial relationships can be built.

Don't just take it from a farmer with dirt in his mouth. Or from an entrepreneur burrowing his way from producer A to consumer B. Or from Gandhi. Or from Muhammad Yunus, recent Nobel Laureate for his micro-lending innovations among the world's poorest. Or, even, from Carlo Petrini, Slow Food philosopher-king.

Take it from Henry Ford. In 1922, Ford wrote presciently about the hazards of prioritizing finance over "natural process":

The most surprising feature of business as it was conducted was the large attention given to finance and the small attention given to service. That seemed to me to be reversing the natural process, which is that the money should come as the result of the work and not before the work. The second feature was the general indifference to better methods of manufacture as long as whatever was done got by and took the money. In other words, an article apparently was not built with reference to how greatly it could serve the public but with reference solely to how much money could be had for it—and without any particular care whether the customer was satisfied. To sell him was enough. A dissatisfied customer was regarded not as a man whose trust had been violated, but either as a nuisance or as a possible source of more money in fixing up the work which ought to have been done correctly in the first place. For instance, in automobiles there was not much concern as to what happened to the car once it had been sold. How much gasoline it used per mile was of no great moment; how much service it actually gave did not matter; and if it broke down and had to have parts replaced, then that was just hard luck for the owner. It was considered good business to sell parts at the highest possible price on the theory that, since the man had already bought the car, he simply had to have the part and would be willing to pay for it.

The automobile business was not on what I would call an honest basis, to say nothing of being, from a manufacturing standpoint, on a scientific basis, but it was not worse than business in general. That was the period, it may

> be remembered, in which many corporations were being floated and financed. . . . The simple idea of doing good work and getting paid for it was supposed to be slow for modern business. The plan at that time most in favor was to start off with the largest possible capitalization and then sell all the stock and all the bonds that could be sold. . . . It was the stocks and bonds, not the work, that mattered.[47]

It might be said that modern finance, given what is known today about the state of the world and the feedback loops of stressed natural systems, is no longer operating in an "honest" and "scientific" way. We continue to invest in a series of corporate and consumer sprints, knowing full well that the outcome of the economic marathon is increasingly in doubt. We continue to apply investment capital as if it were gasoline, as if speed and mobility were the only things that mattered, the only outputs that are real.

"The public thought nothing of a car," Ford wrote, "unless it made speed—unless it beat other racing cars. . . . I never thought anything of racing, but the public refused to consider the automobile in any light other than as a fast toy."

Decades later, with the future of the car company that bears his name in doubt, we are experiencing the limits of the culture of speed. Fascination with the speed of a horseless carriage gave way to addiction to the speed of the highway and to the meta-speed of the electronic superhighway. Now, a growing minority—perhaps not a numerically significant minority yet, but a minority of beyond-nominal significance—is looking to the economy to do more than supply fast toys to those who can afford them. Put another way:

---

47. As cited in *Lapham's Quarterly* (Spring 2008), 69–70.

More people every day are recognizing that the world may not be able to afford ever more, ever faster toys, in the sense that *affording* means more than *having enough fast money with which to purchase them*. More honest, more scientifically justifiable definitions lead us away from fast, faster, fastest, toward values that come before and after those of speed and power.

Consider the formula:

Ford + Yunus + Petrini = Slow Money

Let us ignore for a moment that Henry Ford railed with almost Islamic implication and poetic fury against the idea of arbitrary interest rates and speculative finance.[48] Let us proceed, keeping Ford in mind, to Muhammad Yunus, who promotes the idea of "social business," by which he means for-profit companies whose primary mission is not the creation of shareholder value, but rather the development of market solutions to social problems.

---

48. Ford wrote:

> I have never been able to understand on what theory the original investment of money can be charged against a business. Those men who call themselves financiers say that money is "worth" 6 percent, or 5 percent, or some other percent, and that if a business has one hundred thousand dollars invested in it, the man who made the investment is entitled to charge an interest payment on the money, because, if instead of putting that money into the business he had put it into a savings bank or into certain securities, he could have a certain fixed return. Therefore, they say that a proper charge against the operating expenses of a business is the interest on this money. This idea is at the root of many business failures and most service failures. Money is not worth a particular amount. As money it is not worth anything, for it will do nothing of itself. The only use of money is to buy tools to work with or the product of tools. Therefore money is worth

If well managed, Yunus argues, such businesses would have the capacity to return investors' capital, after which the social business would retain all profits, while the investors retain their ownership and governance rights. The concept of a social business sector, complete with reference to a social stock exchange, is described in *Creating a World Without Poverty*, which presents Yunus' program for complementing the market of enterprises organized around "the profit-maximization principle" with an emerging tier of enterprises organized around "the social-benefit principle."

Who would invest in these social businesses? Yunus suggests that funders will see these enterprises as an attractive alternative to traditional philanthropy. He offers as an example of a social business Grameen Danone, a joint venture between the Grameen Bank, which Yunus founded, and the French yogurt company, Danone. The mission of the company is to improve child nutrition in Bangladesh by producing and marketing yogurt that is

---

what it will help you to produce or buy and no more. If a man thinks that his money will earn 5 percent, or 6 percent, he ought to place it where he can get that return, but money placed in a business is not a charge on the business—or, rather, should not be. It ceases to be money and becomes, or should become, an engine of production, and it is therefore worth what it produces—and not a fixed sum according to some scale that has no bearing upon the particular business in which the money has been placed. Any return should come after it has produced, not before.

What was it that was purchased?
Neither ear of corn nor Henry of Ford,
towards the silo and the showroom blessed,
could say, until this source of Delphic manufacture told:
What to the quality of toil is nothing but a quantity of soil
to the financier is sold as nature's certain interest.

—Edward de Vere, "The Moneylender of Turin"

(There is no theory that connects financial interest, poetry, and soil fertility. But that doesn't mean there shouldn't be.)

affordable on a regular basis to the country's poor. Pursuit of this mission has required many departures from Danone's large-scale manufacturing and marketing experience, including manufacture in "tiny" plants, sourcing milk from "micro-farms," minimal availability of refrigeration, and the use of a locally made molasses from palm-tree dates.

Yunus is convinced that established philanthropic institutions will find such social business to be an attractive asset class: "Once foundations think about social business as a worthwhile target for support, the possibilities begin to seem unlimited. Microcredit can be a very attractive social business. Health care, information technology, renewable energy, environmental remediation, nutrition for the poor, and many other kinds of enterprises can be other arenas for interesting social businesses. Foundations, then, can be a great source of funds for social business."[49] There is a catch, however, and it has proven, over these past few decades, a rather daunting one.

The catch is that within the legal structure and operating cultures of foundations, the cultural divide between "deal doing" and "doing good" is as strong as, or stronger than, anywhere else in the fiduciary universe. Why? Because the investment managers feel entitled to pursue profit-maximizing strategies in the name of "generating more money to give away." And because the language, training, and culture of professional grant makers is very different from those of professional investors, communications and collaboration between these two camps is extremely difficult. As a result, the movement toward mission-related investing by foundations has been glacial, despite such innovative voices as

49. Muhammad Yunus, *Creating a World Without Poverty* (PublicAffairs, New York, 2007).

Jed Emerson, who led the way for many years as president of the Roberts Enterprise Development Fund, developing a well-articulated theory and practice of "blended value" investing. During the 1990s, the Jessie Smith Noyes Foundation also pioneered the use of investment assets to support philanthropic purpose. The F. B. Heron Foundation has been a consistent leader in related explorations, putting innovative mission-related strategies to work in its own shop, as well as seeking to spread this vision within the philanthropic community.

When push comes to shove, however, the foundation community has remained remarkably intransigent in the face of the realization that hundreds of billions of dollars of foundation assets are invested without concern for the foundations' charitable purposes, and, sometimes, in ways whose consequences are the object of grant-making programs, prompting one NGO leader to refer to philanthropy as "the run-off of a broken system."

In 2007, *Los Angeles Times* coverage of the investments of the Bill and Melinda Gates Foundation reminded us just how resistant to change the foundation community can be. The foundation, which is the world's largest, is invested in Nigerian oil drilling operations, whose pollution causes significant health problems in local villages. At the same time, one the foundation's charitable aims is to improve health in Africa. Despite the scrutiny and public dialogue that surrounded the *Los Angeles Times* articles, the foundation "explicitly rejected mission-related investing" in order to "prioritize our program work over ranking companies and issues" and to "stay focused on our core issues."

To those outside of the world of professional philanthropy, such determination is counterintuitive and disheartening.

The search will have to continue for the missing ingredient

that will stir the mind and heart of the modern investor, whether that investor is a fiduciary managing a foundation endowment or an individual investor trying to reckon with the consequences of investing on a smaller scale.

Guilt won't do it. Fear won't do it. A sense of noblesse oblige won't do it. A bit of cognitive dissonance won't do it. These may all be necessary, but they are not, even in the aggregate, sufficient. There are simply too many philosophical cul-de-sacs, too many strategic and organizational dead ends, too many disorienting bifurcations preventing even the well-intentioned investor from leaving the safe haven of profit maximization.

Perhaps the missing ingredient is "Petrini."

Let's revisit the formula:

Ford + Yunus + Petrini = Slow Money

And restate it as follows:

(Non-Speculation) – (Poverty) + (Slow Food) = Slow Money

To Henry Ford's insistence on honesty and scientific integrity over speculative finance and Muhammad Yunus' dedication to social benefit and poverty reduction, we now add Carlo Petrini's vision of the power of food as a tool for social change.

The agenda of Slow Food touches on a great many critical social and environmental issues: soil fertility, nutrition, biodiversity, indigenous culture, artisan food traditions, the joys of the hearth, the value of conviviality. Perhaps Slow Food's greatest contribution comes not from any one of its ends, however, but from its means.

Slow Food is a convener, a community builder, a preserver and restorer of relationships. Through more than a thousand food communities in more than a hundred countries, Slow Food nurtures relationships that are vital to long-term health. In Slow Food's approach, we find something of an antidote to the wounds of the modern economy, a natural remedy to the fever of economic growth. When Slow Food connects producer and consumer, it is leading a process of cultural healing that will ultimately make it possible for investor to reconnect with citizen.

The significance of this process of cultural healing is nowhere more evident than in the sweat-stained cap on the head of one Jack Lazor.

Jack is the farmer-owner of Butterworks Farm, which has been making organic yogurt for more than 30 years on a 325-acre Vermont farm, growing all feed for its dairy herd and developing a sustainable business that is, perhaps, as great a testament to the qualities of slow, small, and local as you will find anywhere. In the taste of his product is an "aliveness," a *terroir* that is remarkable, giving tangible meaning to Carlo Petrini's call for food that is "good, clean and fair." But it is neither to the taste of his yogurt, to its healthy attributes, to the health of his land, to the care of his cows, nor too the remarkable manner in which his small business has grown to $1 million in sales over a few decades, but, rather, to Jack Lazor's cap that we must pay attention now.

"One last question, Jack. What's with the Terra Madre cap? Did you go?"

"Yes, I went to Terra Madre in 2006," he answered.

"What did you think?" his visitor asked, wondering how an

international gathering of 5,000 small farmers in Turin, Italy would relate to the life and work of this Vermonter, whose dedication to the place in which we were standing, to "right here," was so remarkable.

Without hesitating, Jack answered, "It was one of the most meaningful things I've ever done."

Some of that meaning is captured in Carlo Petrini's broad overview of the community of small farmers and slow food enterprises:

> Round the world, no food culture is more important than another. Every single one expresses a profound identity and its language precisely through food.
>
> We have to respect these diversities. We have to be grateful to the art and skill of women and men capable of producing foods as simple as they are outstanding. . . .
>
> Feijoada in Brazil, couscous in the Mediterranean, tamales in Latin America, pakora in India, fufu in Africa, dried reindeer meat in Lapland, pasta in Italy—all foodstuffs representative of the great wisdom of humankind, of subsistence economies and of the never-ending fight against hunger.[50]

Change-as-accelerant has been upon us for the past few centuries. A new kind of change-as-decelerant is emerging, not as a successor, but as a vital complement, enabling farmers around the world to rediscover the ways in which the slow, the small, and the local connect them in an emerging global community.

---

50. Carlo Petrini, "Opening Speech," Terra Madre, October 20, 2004. www.slowfood.com/sloweb/eng/dettaglio.lasso?cod=SW_01368

Our search for the slow, small, and local principles of a new kind of fiduciary responsibility takes us from Ford's insights into speculation to Yunus' conception of social business to Petrini's visceral understanding of food and culture—all the way, it would seem, to the realm of the Earthworm School of Fiduciary Responsibility:

Google + Stonyfield + Butterworks = ESFR

Capital markets have been voting early and often for Google, and, to a far, far lesser extent, for Stonyfield, the world's largest manufacturer of organic yogurt and part of the Danone family. Butterworks Farm might as well be invisible. From a capital-markets standpoint, it simply does not exist. Is the right balance of "economic ingredients" 98.6 percent Google, 1.3 percent Stonyfield, 0.1 percent Butterworks? Should the weighting be shifted from left to right? We cannot say with any precision at the outset of our journey toward slow money. We can say, however, that if the formula does not include Butterworks at all, then capital markets are incomplete, scientifically inaccurate, and, even, to echo Henry Ford further, downright dishonest.

Recognition of this incompleteness is the beginning of the path to a new school of fiduciary responsibility, one in which investors measure their success as much by what they *leave in* the soil for the benefit of future generations as by what they *take out.*

Such a vision of fiduciary responsibility requires holding the reality of Butterworks Farm alongside the reality of hunger in sub-Saharan Africa and food riots in Haiti. It requires holding a calculator in one hand and a compost bucket in the other. The difficulty of doing so is compounded by an investor mind-set that is unable to maintain its balance without ever-greater speed.

It is balance we are after, but not the balance of a gyroscope. We need a more integral form of balance, based not on one-dimensional hyper-speed, but on the stability of multiple rates of change.

In a society that is increasingly held hostage to those aspects of culture that are speeding up, as Steward Brand observes in *Clock of the Long Now*, it becomes increasingly important to protect aspects of culture that "slow way down, just to keep balance":

> The combination of fast and slow components makes the system resilient, along with the way the differently paced parts affect each other. Fast learns, slow remembers. Fast proposes, slow disposes. Fast is discontinuous, slow is continuous. . . . Fast gets all the attention, slow has all the power. . . .
>
> I propose six significant levels of pace and size in the working structure of a robust and adaptable civilization. From fast to slow the levels are: Fashion, Commerce, Infrastructure, Governance, Culture, Nature. In a healthy society each level is allowed to operate at its own pace, safely sustained by the slower levels below and kept invigorated by the livelier levels above. . . .
>
> One of the stresses of our time is the way commerce is being accelerated by global markets and the digital and network revolutions. The proper role of commerce is to both exploit and absorb these shocks, passing some of the

> velocity and wealth on to the development of new infrastructure, at the same time respecting the deeper rhythms of governance and culture.[51]

If we are going to make such understanding real, then we are going to have to make it real in the world of finance. For how can we allow money to accelerate endlessly, hoping that it will not accelerate commerce, erode culture, and degrade nature?

"The rhythms that govern local economies are very similar to our own metabolism," says Carlo Petrini. "The metabolism of nature, the metabolism of man is not fast. It is tranquil. It is slow. The economy must respect this rhythm, because only in that way will the results be positive, strong, and durable."

In the Earthworm School of Fiduciary Responsibility, the notion that a fiduciary's responsibility ends with maximizing shareholder value and minimizing risk seems as silly as suggesting that the pace of commerce is the same as the pace of nature. "If commerce is allowed by governance and culture to push nature at a commercial pace," Stewart Brand concludes, "all-supporting natural forests, fisheries and aquifers will be lost." So, the earthworm-minded investor views it as the highest manifestation of his or her responsibility to ensure that the movement of capital respects and interfaces productively with the slower spheres of culture and nature.

To the earthworm-minded fiduciary, it is clear that philanthropy, venture capital, and government programs—as if locked in vast rows of mechanized, financial monoculture—have proven inca-

51. Stewart Brand, *The Clock of the Long Now* (New York: Basic Books, 1999).

pable of reorienting effectively toward the preservation and restoration of soil fertility and local food communities.

Here are some macroeconomic numbers that never quite seem to be "done."

- In 2007, total grant making by U.S. foundations was approximately $40 billion. Only about $1 billion went to environmental causes, the rest going mostly to churches, schools, hospitals, poverty programs, disaster relief, and the arts. $50 million or so went to organics and sustainable agriculture. At a time when awareness of global environmental threats is sweeping through the general population like a virus, professional philanthropists are allocating only 1 percent of their grant-making budgets to environmental initiatives. And only 0.1 percent to food systems. Meanwhile, the "corpus," as it is called, that is, the $550 billion of investment assets the returns on which fund grant budgets, is invested with virtually no concern for either general issues relating to global economic growth or specific issues relating to a particular foundation's mission.
- In 2007, a total of $20 billion of professionally managed venture capital was invested by venture capitalists in early-stage companies. There were 650 venture capital firms in the United States and a total of $100 billion under management. Approximately $3 billion was invested in 2007 in so-called "clean tech," including solar, wind, and fuel cell technologies. Only a handful of firms, with a total of less than $1 billion under management, included organics in their investment objectives,

with many of these interested more in nutritional supplements and nutraceuticals than in agriculture or organic food.

- In 2007, the U.S. Department of Agriculture had a budget of $92 billion, of which less than $100 million went to small or mid-size organic farmers and less than $10 million went to organic research.
- Approximately 0.5 percent of all agricultural land in the United States is organic. (In the world's most progressive developed nation in this regard, Denmark, the number is 4 percent.)

At a time when there is ample reason to believe that we may be presiding over irreparable damage to the natural systems on which all life depends, these numbers do not represent, in the aggregate, anywhere near an adequate response. We might as well round down to zero. With each passing year, this becomes more of a species-level embarrassment.

While philanthropy, venture capital, and government funding sources are failing to adequately support the emerging sustainable agriculture and organic sector, the diminishing returns of industrial agriculture just keep on diminishing:

- Global pesticide use has increased more than 50-fold since 1950, and most of today's pesticides are more than 10 times as toxic as those used in the 1950s. Less than 1 percent of these pesticides reaches the target pests.
- Global application of nitrogen fertilizer has increased 8-fold since 1960, to more than 80 million metric tons.
- In the United States, the share of the consumer's dollar

that trickles back to the farmer has plunged from 41 cents in 1950 to 19 cents in 2006. Shares to machinery, agrochemicals, seeds, processing, shipping, brokerage, advertising, and retailing have expanded. The USDA reports that the food marketing system grew from $26 billion in 1950 to $717.5 billion in 2006. On the typical Iowa farm, the farmer's profit margin has dropped from 35 percent in 1950 to 9 percent today. In order to generate the same income, this farm would need to be roughly four times as large today as in 1950.

- In Nebraska and Iowa, up to one-third of farmers are in imminent danger of going out of business. Sweden expects to lose half its farmers over the next decade. New Zealand expects to lose 6,000 dairy farms over the next 10 to 15 years, a loss of nearly 40 percent.
- The American flavor industry now has annual revenues of about $1.4 billion. Approximately 10,000 new processed food products are introduced every year in the United States. Almost all of them require flavor additives. Nine out of ten of these new food products fail.
- One government health official compared sanitary conditions in a modern feedlot to "those in a crowded European city during the Middle Ages, when people dumped their chamber pots out the window, raw sewage ran in the streets, and epidemics raged."
- A modern meat-processing plant can produce 800,000 pounds of hamburger a day. A single animal infected with *E. coli* can contaminate 32,000 pounds of ground beef.

- In the Andean highlands, a single farm may include as many as 30 to 40 distinct varieties of potato; in the United States, 4 closely related varieties account for 99 percent of all potato production. Since 1972, the number of U.S. midwestern counties with more than 55 percent of their acreage planted in corn and soybeans has nearly tripled. The 12 million hogs produced by Smithfield Foods, the largest hog producer and processor in the world, are nearly genetically identical.

If we want to restore and preserve soil fertility, if we want to preserve and restore small and midsize farms and promote organic agriculture, if we want to diversify and decentralize our food supply and revitalize local communities, if we want to preserve biodiversity, if we want to remediate polluted and depleted aquifers, if we want to promote human health and childhood nutrition—if we want these and many other related benefits, or even if we merely wish to defensively invest a portion of our assets in a food-system safety net, then we are going to have to figure out how to deploy capital appropriately, in new ways, in meaningful quantities, and for the long term.

What is needed is a new form of financial intermediation. Intermediation that favors a large number of small, independent, local-first food enterprises over a small number of large, consolidated, global-first food enterprises. Intermediation that is oriented around nurture, rather than acceleration, extraction, and export. Intermediation that never loses sight of the difference between wealth and *illth*,[52] and between illth and tilth. Intermediation that is to colossal mutual funds and pension funds what a CSA is to a Wal-Mart. Intermediation that is the opposite of Invisible-Handed, that is, intermediation that seeks to connect investors to that in which they are investing, that seeks to make feedback loops and relationships visible, transparent, healthy. Intermediation whose ultimate goal is to empower investors, entrepreneurs, and farmers as agents of restoration and preservation in their local communities.

Is such a thing possible?

It will require deconstructing typical asset classes and risk/return benchmarks. It will require experimentation along the boundaries of for-profit and not-for-profit. It will require looking at places first, and then at markets and sectors and asset classes.

It will require experimentation with portfolio design: Can the risks of investing in small food enterprises be mitigated by investing in farmland? What about sustainably managed timberland in the region? What are by traditional investment criteria disparate sectors become elements of an integral strategy for investing—or, what Paul Muller calls "reinvesting"—in fertility and health.

We are turning a compost pile. The following design questions are emerging:

- Can we design ways for regional investors to invest in regional food enterprises? Could there ever be Slow Money Bonds, similar to municipal bonds, but investing in local food systems?

52. A distinction, introduced by nineteenth-century cultural critic John Ruskin, that describes wealth created by "progressive industries" as opposed to wealth created by "ruinous chicane."

- Can portfolios of SFEs deliver positive rates of return to investors? Is public or private subsidy required?
- What is the difference between "local" and "regional"?
- If Denmark has a single CSA that is as large as all of those in the United States combined, what does this say about the entrepreneurial opportunities to connect U.S. farmers and U.S. consumers?

Forget the compost pile: This is one moment when the idea of slow decay and regeneration is inappropriate. We are in peril. We are the roadrunner and we have run off the cliff. Suddenly, we must decide: Do we use GMOs and industrial food systems to scramble back to the cliff edge, so that we can resume our journey at full speed along another route? Or do we soften our fall, and increase the chances of our survival, by deploying a slow-money parachute?

False choices, like false dichotomies, are easy to articulate. Either you believe in Adam Smith's Invisible Hand or you don't. Either you believe in the power of free enterprise or you don't. Either you are a real investor, an investor shooting at the target of maximum profits and maximum growth, or you are not. Either you are a true fiduciary or you are a false Malthusian. Either you are with us or you are against us.

The choice between industrial agriculture and slow-money parachute is a false one. We need both. And we can have both—magically transforming ourselves from a roadrunner to a tortoise, that great Native American symbol for the earth itself, and, as we all learned in kindergarten, the winner of the race in the end—by redefining our notions of fiduciary responsibility and prudence.

Our vision of prudence is incomplete. Our ideas of risk and return are incomplete. They are treated as certainties by the modern fiduciary, who needs reminding that, in the words of economic historian R. H. Tawney, "the certainties of one age are the problems of the next."

The certainties of our age? Let's venture a few: Economic growth is the cornerstone of well-being. Speed and power are the solutions of choice. Distant markets trump local markets. Consumerism trumps tribalism.

Beneath such certainties is profound confusion about means and ends:

> Few can contemplate without a sense of exhilaration the splendid achievements of practical energy and technical skill, which, from the latter part of the seventeenth century, were transforming the face of material civilization, and of which England was the daring, if not too scrupulous, pioneer. If, however, economic ambitions are good servants, they are bad masters. Harnessed to a social purpose, they will turn the mill and grind the corn. But the question, to what end the wheels revolve, still remains; and on that question the naïve and uncritical worship of economic power, which is the mood of unreason too often engendered in those whom that new Leviathan has hypnotized by its spell, throws no light. Its result is not seldom a world in which men command a mechanism that they cannot fully use.[5]

53. R. H. Tawney, *Religion and the Rise of Capitalism* (London: John Murray, 1944).

It is toward the "full use" of the power of entrepreneurship that we want to head, as we first bring multiple bottom lines to the entrepreneurial equation, and then hope to get above and below all of them.

Tom Miller, the first CEO of the Kentucky Highlands Corporation and later head of Program Related Investing at the Ford Foundation, offers a glimpse of the way toward "completion," beyond conflicting worldviews and false choices:

> We are not trying to reign in or correct or punish capitalism. We are trying to complete it.
>
> Capitalism remained incomplete because resources seemed inexhaustible and consumption seemed to cause no harm. Now, as we reach a new juncture in our history on the planet, this is no longer the case.
>
> Our most fundamental resources are in large part not counted in the costs of products. Sure I pay for water, but that accounts only for the costs of getting the bugs out of it and getting it to my tap. There is no accounting for the costs of depleting an aquifer. Sure I pay for gasoline, but that accounts only for the costs of getting it from the oil patch to my gas tank. There is no accounting for what happens to the air and the planet after I burn it. Sure I pay for lima beans, but their price only accounts for the costs of planting and growing and delivering them to my nearest Wal-Mart. There is no accounting for the costs of depleting soil nutrients or harm to rivers by fertilizer.
>
> This is not a failure of capitalism, but merely the result of an incomplete system, which it is up to us to complete.[54]

54. Miller, e-mail message to author, April 27, 2008.

Miller's observations about incompleteness highlight a curious development. Despite Paul Samuelson's observation that in the twentieth century mathematics swept through the field of economics like a virus, economics does not seem to have recognized the idea of incompleteness, which is one of the cornerstones of modern mathematical thinking.

It was Kurt Gödel, a colleague of Albert Einstein's at Princeton, who developed the incompleteness theorem. Philosophy professor Rebecca Goldstein ranks it, along with Einstein's theory of relativity and Heisenberg's uncertainty principle, as part of "that tripod of theoretical cataclysms that have been felt to force disturbances deep down in the foundations of the 'exact sciences.'"[55]

Gödel's theorem states:

> To every ω-consistent recursive class κ of *formulae* there correspond recursive *class-signs* r, such that neither v Gen r nor Neg (v Gen r) belongs to Flg (κ) (where v is the *free variable* of r).

Or, in conversational English:

> All consistent axiomatic formulations of number theory include undecidable propositions.[56]

Or, as taught in high school:

---

55. Rebecca Goldstein, *Incompleteness: The Proof and Paradox of Kurt Gödel*, (New York: W.W. Norton & Co., 2005).

56. Douglas Hofstadter, *Gödel, Escher, Bach: An Eternal Golden Braid* (New York: Basic Books, 1979).

> Every formula is either complete and inconsistent, or consistent and incomplete.

That is, mathematical formulas will work only if you exclude certain inconsistencies. This may be akin to the problems of applying Euclidean geometry, which works perfectly on a plane, to the real world, which is curved.

Goldstein observes that the incompleteness theorem is more than just mathematical—it is "metamathematical:"[57]

> Gödel's theorems, then, appear to be that rarest of rare creatures: mathematical truths that address themselves—however ambiguously and controversially—to the central question of the humanities: what is involved in our being human. They are the most prolix theorems in the history of mathematics. . . . What they say extends beyond mathematics, certainly into metamathematics and perhaps even beyond.
>
> The prefix *meta* comes from the Greek, and it means "after," "beyond," suggesting the view from outside, as it were. . . .
>
> Einstein's and Gödel's meta-convictions were addressed to the question of whether their respective fields are descriptions of objective reality—existing independent of our thinking of it—or, rather, are subjective human projections, socially shared intellectual constructs.[58]

---

57. Note that *Webster's Third* shows *metamathematics* as not hyphenated. *Meta-economics,* with or without a hyphen, is not found.
58. *Incompleteness.*

The idea of the bottom line is a socially shared intellectual construct. The idea of the triple-bottom-line is an intellectual construct that is beginning to be shared more widely. We are on the threshold of new constructs of fiduciary responsibility and prudence that are capable of redefining the relationship between markets and places.

No exploration of slow money written at this moment in time could fail to include the words "Bear Stearns," the preservation of which by the Federal Reserve and JPMorgan Chase was an early finger in the dyke of the global credit crisis of 2007 and 2008.

We may leave to experts guesswork about money supply, funds rates, capital reserves, recession, and regulatory schemes that might avert or dampen the effects of the excessive speculation that haunts—no, is built into—global financial markets. When former secretary of the treasury Robert Rubin says, "No one knows," we may take him at his word, and marvel at the extent to which the world rushes, nevertheless, after the counsel of the very investment bankers who aided and abetted the impenetrable complexity and financial instability that is now our problem. (This is a corollary of the irrational fiduciary mind-set that drives investors to pay management fees to money managers, despite the overwhelming evidence that the vast majority of money managers underperform the market, particularly net of those very fees.)

Let's leave the general story of the Bear Stearns debacle now for a story of a more personal nature.

We are sitting in the harborside home of a retired executive of Bear Stearns. Joining him is the executive director of a nonprofit organization that is promoting sustainable development in the historic coastal community in which both live. After considerable conversation, covering such topics as soaring home prices, lack of affordable housing, explosive real estate development and speculation, expanding transportation infrastructure, and the collapse of the scallop fishery, the retired executive says:

> If you are telling me that there is a pipe that is dumping toxic effluent into the harbor and killing the scallops, I will be the first person at the head of the line to shut that pipe down. But if you are telling me that there are too many houses, too many lawns, too many septic systems, too many cars, too many boats, and that it is the cumulative effect of all of these, well, I call that progress, and if losing the scallops is the price of progress, then that is a price I am willing to pay.

The idea of economic violence was nowhere in his field of vision. The fact that Thomas Malthus' mug had recently appeared in the *Wall Street Journal* seemed irrelevant.[59]

It was as if George Soros were whispering in his ear: "The financial market is amoral in that respect, because individual investors

---

59. The lead article in the *Wall Street Journal* on March 24, 2008, was titled "New Limits to Growth Revive Malthusian Fears." The article concludes: "History is littered with examples of societies believed to have suffered Malthusian crises: the Mayans of Central America, the Anasazi of the U.S. Southwest, and the people of Easter Island. . . . Those societies, of course, lacked modern science and technology. Still, their inability to overcome resource challenges demonstrates the perils of blithely believing things will work out, says economist James Brander. . . . 'We need to look seriously at the numbers and say: Look, given what we're consuming now, given what we know about economic incentives, given what we know about price signals, what is actually plausible?' says Mr. Brander."

can't affect the outcome. And that's a very happy position to be in, because then I don't have that moral problem. . . ."[60]

None of us is immune to the inexorable tendency of finance to separate heart and mind.

A few years ago, after listening to investment presentations from entrepreneurs, a member of Investors' Circle said from the podium to a hundred or so of his fellow investors, "I don't know about you, but I wish I could write a check for $500,000 to most of the companies we just heard from."

Responding to the dead silence that greeted this remark, he added, "I realize that would make me a crappy venture capitalist."

Laughter erupted.

That laughter arose from the territory between heart and mind. Their venture capital minds knew that they had to say no to 99 out of 100 ventures if they wanted to construct a portfolio that had a chance of delivering a 20 percent IRR. Their engaged-citizen hearts knew that the world was a better place because of those entrepreneurs. Somewhere in between, they wanted to reduce the start-up financing risks of those companies, to get them out in the marketplace more quickly, knowing full well that most of them would likely fail.

Their laughter came from a place of clarity that lay just beyond reach.

There is a word for what comes after or lies within such laughter. It is *epiphany*.

---

60. Interview with Terry Gross on National Public Radio's *Fresh Air*, July 13, 2006.

Shareholder advocacy will not be sufficient. Double-bottom-line or triple-bottom-line social investment portfolios will not be sufficient. Investing in clean tech and renewable energy will not be sufficient. Ending subsidies for oil companies and commodity farms won't do it. Dragging 1 or 2 percent of foundation assets across the line into mission-related investing won't do it. If we are going to have a chance of steering the economy in a fundamentally more restorative direction over the next generation, then we are going to have to learn a new way to see our place in the economic scheme of things and the place of the economy in the natural scheme of things.

We will need to head, impelled by sudden and irrepressible insight, toward beauty and nonviolence.

Beauty seems an unexpected destination in the journey toward a new way of investing, until we recognize that what comes after the beautiful humanism of the first Renaissance might be an even more beautiful *species-ism* of the second Renaissance, reconnecting us to the world in new, meta-economic ways.

Toward this end, we should appreciate and nurture the beauty of small farms and diversified cultivation. There is something beautiful in CSAs. There is something beautiful about organic garlic grown in Dixon and sold in Santa Fe. There is something beautiful about Davis Farm eggs at the Guerilla Café. There is something beautiful about Bayley Hazen Blue Cheese from Jasper Hill Farm. There is something beautiful in the manure between the rows of *Fascianella* almond trees near Noto. There is something beautiful in the Nebbiolo from Castello di Verduno. There is something beautiful in the *lumache* of Cherasco. There is something beautiful in the fermented koumiss at the market in Osh. There is something beautiful in the green cumin and *zibibbo* grapes of Khorasan. There is something beautiful in *cochayuyo*, harvested from the sea by the women of Pichilemu. There is something beautiful in the *bokhan* bread of Kobrin. There is something beautiful in the prickly pears of Tlaxco. There is something beautiful in a Mora Romagnola pig.[61]

There is nothing beautiful about bovine growth hormone or high-fructose corn syrup or Red Dye # 4 or the parking lot of a McDonald's.

Or maybe we should rest content with the clarity of laughter:

> A rabbi, a Hindu priest, and a hedge-fund manager were lost one night in Kansas. In the distance, they saw a light on in a farmhouse, drove up, and knocked on the door.
>
> "Excuse us," the priest explained when the farmer came to the door. "We know it's late. But we are lost and we're too tired to continue on. By any chance could we stay here for the night? We won't be any trouble, and we'll leave first thing in the morning."
>
> "Happy to help," the farmer said. "But there's only one problem. I only have two spare bedrooms. One of you will have to sleep in the barn."

61. "The Mora Romagnola grows much more slowly that the omnipresent Large White and its meat has more fat content than other breeds, which is why it was completely abandoned with the spread of commercial breeding and is now at risk of extinction. The nearly 22,000 head that existed in 1949 were reduced to just 12 in 1997, miraculously preserved by Mario Lazzari, an old breeder from Faenza. The animal's tasty, tender, but compact meat is perfect for making excellent cured meats." From *Terra Madre 2006* (Bra, Italy: Slow Food Editore, 2006).

"Oh, I'm happy to stay in the barn," the priest replied immediately. "No bother at all."

They all bed down for the night, but after a few minutes, there was a knock at the door. It was the priest.

"You didn't tell me you had a cow in the barn," he said. "It's not exactly against my religion to sleep near a cow, but I don't quite feel comfortable."

"No problem," the rabbi jumped in. "Don't worry. I've got no problem with the cow. You take my room. I'll sleep in the barn."

And so they went to bed again, but, lo and behold, in a few minutes, there was a knock at the door. It was the rabbi.

"You didn't tell me you had a pig in the barn," the rabbi explained. "I'm awfully sorry, but it's like with the priest. There's no real religious issue with my sleeping with pigs, but it still makes me uncomfortable."

"Jesus Christ," the hedge-fund manager said, losing all patience. "I won't have any religious issues with the animals! I'll sleep in the barn. Come on, let's all get some sleep!"

So the priest and the rabbi went upstairs with the farmer, and everyone went to bed, confident that things had been taken care of once and for all.

But in a few minutes, there was a knock at the door.

It was the cow and the pig.

THREE

back down to earth

It is good, it is appropriate that we use the left side of our brain to study markets and industry segments and capital flows, to measure risk and return, to mete out liquidity and diversification and various instruments designed to intermediate efficiently between producers, consumers, and the natural systems on which all life depends.

But it is even better, at this moment in history, to use the right side of our brain, and our whole heart, and whatever portion of our spirit can be brought to the task, to take the first steps, new steps, imperfectly charted steps, toward the realm of slow money.

We need to steer money, in its primary applications, that is, in those functions toward which it is deployed in *the making of money*, toward life, toward enterprises that enhance the quality of life, that preserve and restore fertility, biodiversity, and the health of bioregions and communities and the households that live in them, and away from enterprises that degrade quality in the name of quantity. This use of money, of investment capital as an antidote to the disease of excessive quantification is, in the words of a veteran McKinsey consultant, "tricky."

There is something of financial homeopathy in it. A drop of

slow money under the tongue of the body economic: What will its effect be on the health of the whole system? We cannot know, but we can assert and affirm our hope, our intimation that its effect will be salutary.

"We need," in the words of former Investors' Circle member Friedmann Schwarzkopf, "to bring money back down to earth."

We need to stop thinking about money as lubrication for a machine that is everywhere and nowhere and at no given moment, and to start thinking about money as irrigation for the field of our intentions, which are expressed right here, right now, where we live and where we work. We need to stop giving priority to the imperatives of money's explosive self-propagation and start giving priority to the imperatives of social implosion and impending ecological collapse.

Much of what the social investment initiatives of recent years are aiming at can be more directly, more fundamentally understood as a problem of the speed of money. Screened portfolios and shareholder advocacy work to heal the wounds caused by globalization and industrialization and corporatization. As critical as these means of redress are, they remain, to some extent, an exercise in "wake management"—most of their benefits are achieved not by slowing the economic speedboat down, but rather by minimizing some of its impacts as it speeds through the harbor.

Environmental degradation, a throwaway consumer culture, cheapened food (rich in empty calories and chemical additives), media programmers who live by ratings, nightly news reports that cover daily fluctuations of market indices—these are inevitable by-products of an economy whose decision making is driven first and foremost by the imperatives of financial markets, an economy in which money, unleashed through the power of technology and unfettered by either connection to place or the human face of exchange, has taken on a life of its own, a speed of its own.

Fast money does violence to the web of relations on which the health of communities and bioregions depends.

It is not enough to steer money in new directions. We must slow money down.

We are surrounded by the explosive creation of wealth that drives venture capital.

After its IPO in 2005, Google's stock price shot up from $85 to over $700 in eighteen months, an increase in market capitalization from $25 billion to more than $200 billion. Google is the prototypical venture capital deal, the epitome of a process that bets billions of dollars per year on a few thousand technology companies. It is a symbol of virtually limitless upside. A Google search weighs nothing, is silent, and has, to its user, no immediate ecological footprint or cost—an apparently perfect manifestation of the Invisible Hand at work.[1] Although it is possible to imagine a fiduciary asking, "How many McDonald's are too many? Can the world sustain 50,000 McDonald's and a trillion hamburgers?"[2] it is impossible to imagine a fiduciary ever asking, "Would a trillion Google searches a day be 'too many?'" Google is a portal

1. This conjures up Scott Nearing's observation: "To the capitalist, the perfect product costs a dime to make, sells for a dollar, wears out quickly and leaves a habit behind."

2. There are approximately 31,000 McDonald's worldwide, with over 100 billion burgers served to date. There are approximately 14,000 McDonald's in the United States and fewer than 1,000 in China. "McD's Preps for China Drive-Thru Boom" reports the

to the world of unlimited upside, the world of unlimited information and entertainment, the world of unlimited shareholder entitlement.

It is deeply disturbing to stand at the edges of such extreme wealth, such extreme speculation—even when successful—and peer into the expanses of such unrelenting poverty: poverty of abandoned building and abandoned village and field abandoned to mall, poverty of slum and ghetto, poverty of pollution, poverty of congestion and sprawl, poverty of cheapness and impermanence, poverty of gated community and security system, poverty as if ordained by an invisible hand, poverty of the devalued and the overvalued, poverty of entire populations who produce little but consume much, poverty of the near and the real overtaken by the distant and the virtual, poverty of empty calorie and long shelf life, poverty of plastic, poverty of divorce and displacement, poverty of erosion, poverty of proliferating portfolios, poverty of market mania, poverty of irrational exuberance, poverty of affluence.

Our ability to redress this poverty in the twenty-first century will depend on our ability to look beyond wealth creation of the venture capital kind, to look across the boundary of for-profit and nonprofit, to look past IRR and GDP and market indices, and to discover more integral, truer, more beautiful measures of progress and well-being.

With appropriate vision, we will come to see each transaction, each investment not only as a tiny moment of truth, but also as a small but critical opportunity to choose beauty over convenience,

---

business press. There are 15 million cars in China, with government projections pointing to 140 million by 2020. It takes seven pounds of grain to create a pound of beef, and 1,000 gallons of water (or, perhaps as much as 2,500 gallons: this figure is the source of some contention).

beauty over competitiveness, beauty over uniformity, beauty over control, beauty over making a killing, beauty over caveat emptor, beauty over commodification, beauty over the lowest-common-denominator, beauty over buying low and selling high.

We cannot achieve such vision without what economist Herman Daly has called a "deep philosophical clarification."[3] In these times, such a clarification can only be arrived at, finally, and to good effect, through persistent exploration of new ways to integrate values of beauty and nonviolence into economic transactions.

Experiments in slow money are experiments in beauty and nonviolence.

Beauty was good enough for the title of E. F. Schumacher's seminal work, and it should be good enough for us. He could have chosen *Small Is Appropriate*, or *Small Is Good*, or *Small Is the Key to Health and Happiness*. He chose the word *beautiful*; however, his attention focused on issues of scale, nonviolence, self-sufficiency and a "meta-economic" understanding of man's place in the creation: "Divergent problems, as it were, force man to strain himself to a level above himself; they demand, and thus provoke the supply of, forces from a higher level, thus bringing love, beauty, good-

3. "This change in vision involves replacing the economic norm of quantitative expansion (growth) with that of qualitative improvement (development) as the path of future progress. This shift is resisted by most economic and political institutions, which are founded on traditional quantitative growth and legitimately fear its replacement by something as subtle and challenging as qualitative development. The economics of development without—and beyond—growth needs to be worked out much more fully. There are enormous forces of denial aligned against this necessary shift in vision and analytic effort, and to overcome these forces requires a deep philosophical clarification, even religious renewal." Herman Daly, *Beyond Growth* (Boston: Beacon Press, 1996).

ness, and truth into our lives. It is only with the help of these higher forces that the opposites can be reconciled in the living situation."[4]

The unrelenting exercise of the calculus of economics has facilitated the slide toward one-dimensional management decision making, ugly commercial landscapes, horribly littered media spaces, and dumbed-down public discourse. We have noted, and debated with varying degrees of heat, the Death of God. Some have even noted what they have called the Death of Money, as currency shifted from gold to paper to bits of information. It is the Death of Beauty that concerns us here.

If we are to rediscover meta-economics, then it will be found through a reaffirmation of the primacy of nonviolence and beauty.

In all our actions, especially those that relate to the management of wealth and the making of the enterprises and the products that occupy, and define, so much of our waking hours, we must keep our attention on nonviolence and beauty.

Products produced cheaply create ugly work lives and ugly households and ugly communities. Profits produced quickly cannot purchase patience and care. Patience is beautiful. Restraint and care are beautiful. Peace is beautiful. A small, diversified organic farm is beautiful.

There is nothing beautiful in the idea that we will only do no harm if we can, in so doing, make as much money as is generated in the doing of harm.

4. *Small Is Beautiful.*

Pioneering companies like Stonyfield Farm seek to capture increasing amounts of Wal-Mart shelf space for organics and responsibly produced consumer goods. Pioneering venture funds like Green-mont Capital provide venture capital to organic food companies.

These important enterprises redirect capital away from destructive industrial activity and toward restorative economic activity. However, they do not influence directly the speed of money or redefine directly the role of investors in the evolution of capital markets.

Despite the commercial success of a number of small socially responsible companies, such as Ben & Jerry's, Aveda, and Stonyfield, along with the dramatic growth of the organics and LOHAS markets, as evidenced by Whole Foods' rapid approach to the threshold of the Fortune 500, fundamental questions remain unanswered:

- How is mission inexorably compromised when a company goes public or is acquired?
- What happens to "local" when a company scales?
- Is there an alternative to Series A, Series B, Series C, and IPO?[5]
- Can alternatives to the traditional corporate charter be designed, creating ownership and governance structures that embed a "stakeholder accountability" culture that cannot be diluted as a company grows?
- Would the food system, in particular, and the economy, as a whole, be safer and healthier if tens of thousands of small, independent, mission-driven companies were supported by capital that prioritized local control?

5. This refers to private financings that precede and lead to an initial public offering.

These interrelated questions go to the fundamental imperatives of an economic transformation that started with the Jeffersonian ideal of a small farmer and ends with Twinkies, TV dinners, nutraceuticals, and a bunch of eco-farmers who don't know whether they are a movement or an industry.

Investors who invest in organic food companies without addressing these fundamental questions are too much like organic gardeners who use organic fertilizers and organic pest control, but who know nothing of the joys of composting and the mysteries of soil health. The result is a system that produces organic food companies and puts organic food products on the shelves, but leaves the soil of the community less fertile, less full of life, and less capable of supporting future generations.

Such system-design questions are difficult to address in the macro. We would do well, therefore, to trust a good measure of our future, if only a small portion of our portfolios, to the disaggregated, fertile, diverse creative genius of entrepreneurs and farmers.

Entrepreneurs and farmers are the poets of the economy. They are holders of ambiguity and risk. They cultivate interstitial spaces, where demand and need and aspiration coexist in a mildly turbulent state of chaotic possibility. They continuously test the boundaries of quality and quantity, as a poet tests the boundaries of denotation and connotation. Ideas in a business plan; seeds in potting soil; rhymes in search of new reasons.

Some of which may have been on his mind when Greg Steltenpohl, the founder of Odwalla, said during a recent conversation, "We want to make entrepreneurs more like farmers, not make farmers more like entrepreneurs." Such a statement may have sounded downright daffy a generation or two ago, when mass production and shareholder capitalism were in their infancy. Today, however, we need to reexamine the conventions of entrepreneurship and entrepreneurial finance, taking care not to continue acting in ways that once made sense, but now seem awfully mechanical, awfully prosaic, for a world whose social fabric is fraying and whose biological integrity is under threat.

The rules and objectives of the Industrial Revolution and industrial finance, which have made possible everything from space travel and Velcro to the Internet and Pizza Hut, cannot be extended forever in a straight line. One kind of corporation is not going to serve all the needs of an evolving planet. One kind of capital market is not going to be responsive to the needs of all communities and bioregions. One definition of profit is not going to be appropriate for all stages of economic maturity.

In fact, we might say that the urge to simplify and its paradoxical cousins, befuddlement and monoculture, are culprits that have led modern economic man astray. We have cut everything up into tiny buy-low/sell-high pieces. We have sliced and diced risk into a zillion securitized fragments. We have traded baguettes for Twinkies. We have traded freshness for shelf life.

Contemplating an apology for the extent that such explorations diverge from investment prospectus and strategic planning memorandum, we may remember Pascal's famous apology, "I am sorry that this letter is so long. I did not have time to make it shorter." To which the Thoreauvian fiduciary might add, today: "I'm sorry that I made your money grow so slowly. There is no time to make it grow faster."

We do not have time, any more, to rely solely on fast money.

Fast money made sense when corporations were small and the world was big, when resources and places to dispose of waste seemed infinite, when mass production was first being tapped to fuel higher standards of living. The Prudent Man rule made sense when 100 million shares was a big day on Wall Street and the idea of a $350 billion Wal-Mart or of a $200 billion CalPERS (California Public Employees' Retirement System) or of a venture capitalist whose fees are calculated as percentages of billions of dollars would have seemed preposterous. Buy Low/Sell High made sense as a cornerstone of fiscal responsibility when the Depression and World War II still cast their shadows over the soul of America, and building suburbs and building superhighways and building cars and building skyscrapers and building stockpiles were understood to be synonymous with building national character. The Invisible Hand provided a useful mythology for an age that was moving away from monarchy and toward democracy and had yet to imagine child labor laws or the minimum wage. Wealth Now/Philanthropy Later made sense in the era before the terms *ecology* and *bioregion* and *smog* and *ozone hole* and *eutrophication* and *Evian* and *Desert Storm* and *Farm Aid* and *McWorld* and *organic* and *ultrapasteurized* had entered the vernacular.

We do not have time, any more, to chase speed. We must find new ways to mark our progress.

We do not have time, any more, to rest comfortably in the false securities of specialization. We must let the heart of the generalist lead us back to our proper place in the wider world.

We do not have time, any more, to limit our discourse to the lexicon of unlimited economic growth.

We must not be apologetic for a non-prospectus that is so philosophical. We do not have time to make it more pecuniary.

Slow money is not for venture capitalists. It will not even be for those patient capitalists whose convictions remain centered in the creation of a socially responsible venture capital marketplace that can compete for market share with traditional venture capital.

Slow money is for nurture capitalists.

Behind Eliot Coleman's statement, "At 1,000:1, in four months, a tomato seed makes even the highest fliers seem paltry," is a spirit that is seeking to learn from nature, to align human activity with the workings, and the mystery, of natural systems.

We may wish to call nurture capitalists those investors who take to heart the voices of Wendell Berry, Eliot Coleman, Joan Gussow, Fred Kirschenmann, and Gene Logsdon.

WENDELL: Let me outline as briefly as I can what seem to me the characteristics of these opposite kinds of mind. I conceive a strip miner to be a model exploiter, and as a model nurturer I take the old-fashioned idea or ideal of a farmer. The exploiter is a specialist, an expert; the nurturer is not. The standard of the exploiter is efficiency; the standard of the nurturer is care. The exploiter's goal is money, profit; the nurturer's goal is health—his land's health, his own, his family's, his community's, his country's. Whereas the exploiter asks of a piece of land only how much and how quickly it can be made to produce, the nurturer asks a question that is much more complex and difficult: What is its carrying capacity? (That is: How much can be taken from it without diminishing it? What can it produce *dependably* for an indefinite time?) The exploiter wishes to earn as much as possible by as little work as possible; the nurturer expects, certainly, to have a decent living from his work, but his characteristic wish is to work *as well* as possible. The competence of the exploiter is in organization; that of the nurturer is in order—a human order, that is, that accommodates itself both to other order and to mystery. The exploiter typically serves an institution or organization; the nurturer serves land, household, community, place. The exploiter thinks in terms of numbers, quantities, "hard facts"; the nurturer in terms of character, condition, quality, kind.

It seems likely that all the "movements" of recent years have been representing various claims that nurture has to make against exploitation.[6]

ELIOT: Deep-organic farmers, after rejecting agricultural chemicals, look for better ways to farm. Inspired by the elegance of nature's systems, they try to mimic the patterns of the natural world's soil-plant economy. They use freely available natural soil foods like deep-rooting legumes, green manures, and composts to correct the causes of an infertile soil by establishing a vigorous soil life. They acknowledge that the underlying cause of pest

6. Wendell Berry, *The Unsettling of America* (San Francisco: Sierra Club Books, 1977).

problems (insects and diseases) is plant stress; they know they can avoid pest problems by managing soil tilth, nutrient balance, organic matter content, water drainage, air flow, crop rotations, varietal selection, and other factors to reduce plant stress. In so doing, deep-organic farmers free themselves from the need to purchase fertilizers and pest-control products from the industrial supply network—the commercial network that normally puts profits in the pockets of middlemen and puts family farms on the auction block. The goal of deep-organic farming is to grow the most nutritious food possible and to respect the primacy of a healthy planet. Needless to say, the industrial agricultural establishment sees this approach as a threat to the status quo since it is not an easy system for outsiders to quantify, to control, and to profit from. Shallow-organic farmers, on the other hand, after rejecting agricultural chemicals, look for quick-fix inputs. Trapped in a belief that the natural world is inadequate, they end up mimicking the patterns of chemical agriculture. They use bagged or bottled organic fertilizers in order to supply nutrients that temporarily treat the symptoms of an infertile soil. They treat the symptoms of plant stress—insect and disease problems—by arming themselves with the latest natural organic weapons. In so doing, shallow-organic farmers continue to deliver themselves into the control of an industrial supply network that is only too happy to sell them expensive symptom treatments. The goal of shallow-organic farming is merely to follow the approved guidelines and respect the primacy of international commerce. The industrial agricultural establishment looks on shallow-organic farming as an acceptable variation of chemical agribusiness since it is an easy system for the industry to quantify, to control, and to profit from in the same ways it has done with chemical farming. Shallow-organic farming sustains the dependence of farmers on middlemen and fertilizer suppliers. Today, major agribusinesses are creating massive shallow-organic operations, and these can be as hard on the family farm as chemical farming ever was.

The difference in approach is a difference in life views. The shallow view regards the natural world as consisting of mostly inadequate, usually malevolent systems that must be modified and improved. The deep-organic view understands that the natural world consists of impeccably designed, smooth-functioning systems that must be studied and nurtured. The deep-organic pioneers learned that farming in partnership with the natural processes of soil organisms also makes allowance for the unknowns. The living systems of a truly fertile soil contain all sorts of yet-to-be discovered benefits for plants—and consequently for livestock and the humans who consume them. These are benefits we don't even know how to test for because we are unaware of their mechanism, yet deep-organic farmers are aware of them every day in the improved vigor of their crops and livestock. This practical experience of farmers is unacceptable to scientists, who disparagingly call it mere "anecdotal evidence."[7]

7. www.rakemag.com, September, 2004.

JOAN: Where food regulation is concerned, quantification sets a trap. Before I retired from the University, there had been a number of efforts at the federal or state level to pass laws that would allow only "nutritious" foods to be sold in school vending machines, or during school lunch. Since quantitative guidelines are usually a necessity when laws are written, I used to require students in my nutrition-policy seminar to write a legal (quantitative) definition of a nutritious food. They quickly discovered that by many quantitative standards, an apple isn't nutritious—or it wasn't, until the "authorities" decided complex carbohydrate and water "counted"—so the students used to bend over backwards trying to write some sort of definition that would let an apple in and keep a candy bar out.

How might we draw a line between two products made with organic wheat, a pasta (84% carbohydrate—mostly complex—and some vitamins and minerals) which most of us would probably want to allow, and a Twinkie (65% carbohydrate—mostly simple) fortified to contain significant amounts of various nutrients? In an era when the laxative Metamucil has claimed it contains as much fiber as two bowls of oatmeal, or the table sweetener Equal has tried to position itself "as a healthy alternative" like "2% milk," nutrients alone cannot define a healthy food. What we mean by "nutritious" is something more. We mean, I think, something like wholesome.[8]

8. Joan Dye Gussow, "Can an Organic Twinkie Be Certified?" in *For All Generations: Making World Agriculture More Sustainable*, edited by J. Patrick Madden & Scott Chaplowe (West Hollywood, California: World Sustainable Agriculture Association, 1997). It is interesting to note, given her observation about the concept of "wholesome," that the

FRED: What's happened as a result of the direction that we've gone is that our food system has become polarized. We have direct markets at one end and highly consolidated commodity markets at the other end. To compete in these commodity markets, farmers have to get bigger. There isn't anything in the middle, and the middle is what has traditionally been at the heart of much of agriculture. It has been the independent family farm, farms of 2 to 300 acres. There is an enormous opportunity right now to create networks of mid-sized farmers, similar to Niman Ranch, that produce unique products for niche markets.

GENE: The soil's organic matter is down to an average of 1.5 percent in Ohio, when it should be at least 3 percent for efficient use of fertilizers, tractor power, and moisture-holding capacity, not to mention many other subtle influences on soil health, and therefore food quality. There are, for example, growing critical shortages of trace elements in Ohio soils—chiefly selenium, but also manganese, zinc, boron, and others, shortages that with less intensive farming and with maintenance of the soil's organic matter, were not previously problems. . . . But biological degradation stemming from economic pressure does not stop with the soil. All living things are affected, not the least of which are the farmers themselves and the communities in which they live.[9]

Venture Capital Journal featured a 2008 cover story about socially responsible venture capital investing entitled "Wholesome Investing."

9. *At Nature's Pace.*

The task of the nurture capitalist is to integrate such knowledge about fertility, farming, and food into new, deeper definitions of fiduciary responsibility and entrepreneurship.

Again we must ask: How do we put such a vision into practice?

First, we must be in the business of convening and connecting, much in the way that Investors' Circle, the national network that is incubating Slow Money, convenes and connects investors, entrepreneurs, and other stakeholders, building social capital that catalyzes the flow of financial capital.

Second, we must not be afraid to define a new asset class, something that might be thought of as a "sub-asset" class of patient capital or venture capital.

Third, management practices must reflect a commitment to principles of "appropriate finance," through mechanisms that balance IRR to investors with "ERR" to stakeholders, perhaps through capped returns for investors, with returns above a certain threshold shared by the investors and other stakeholders.

What we are after is not the creation of a fund-management entity that will benefit a tiny group of general partners and a slightly larger group of institutional limited partners, but rather a permanent intermediary organization that can build capacity over time and benefit generations of small investors, agricultural producers, and food consumers.

From the standpoint of investment strategy and portfolio management, we must explore combining direct investments in early-stage food enterprises with a high degree of diversification, by looking at sustainable timber and farmland as asset classes that offer lower risk, while generating substantial benefits in terms of soil fertility and bioregional health. Timber investing is typically managed by large-scale, institutional timber-investment management organizations ("TIMOs"). The Forestland Group is the largest such organization that is dedicated to sustainable practices. Lyme Timber in New Hampshire is far smaller, and focuses on conservation-oriented land investing. The Pacific Forest Trust is a relatively new firm established by former Investors' Circle member Connie Best; the firm is small, but has been quite effective at designing and implementing "ecosystem services" easements that have the potential to enhance financial returns from sustainably managed forests.

A capital market for investing in small organic farms does not exist. More than one million acres of farmland are lost each year to development. Land Trusts typically focus on wilderness and open space, and those that do buy development rights from farmers are usually unable to take ownership of the land. In collaboration with sustainable-agriculture NGOs, it may be possible to manage an investment activity that would typically be impossible for a "venture capital fund": Buy small farms that are on the market, identify new farmers who want to farm organically, lease the farm to them for three to five years, and then sell them the farm, with the development rights purchased by a Land Trust.

Timber and farmland investing of the sort outlined above have management imperatives that are quite distinct from those of investing in portfolios of early-stage companies. Seen through the sector lens of a traditional fiduciary, the inclusion of all three types of investments in a single fund is confusing and impractical. Seen through the "place" lens of slow money, this approach to investing that targets fertility and food systems has the potential to be integral, differentiated, and effective.

Connie Best has asked, "What are sustainable returns?"

Although she is referring to the returns that can be generated from appropriately managed forests, the concept applies more broadly to the fundamental integration of strategies designed to balance shareholder returns with the imperatives of natural systems. In recent years, sustainable timber management has yielded returns in the 6 to 9 percent range.

A Harvard/McKinsey study of $72 million of Investors' Circle investments in 110 deals indicated hypothetical portfolio returns of 5 to 14 percent. Investors' Circle members have invested approximately $24 million in 37 food companies over the past 15 years; however, no aggregated return data is available for this "portfolio," since investments were not managed collectively and were held for varying periods. Stonyfield Farm might be considered an example of a home run in organic food investing: Many of its investors achieved IRRs in the 20 percent range during the company's growth, which led to its acquisition by Group Danone. Other organic companies of roughly similar scale that have created liquidity and attractive returns for investors through acquisition by multinational corporations are White Wave Foods and Horizon Organic.

Our first model of a hypothetical Slow Money portfolio shows returns of 5 to 8 percent. We must be careful to say that this is only a first approximation. We must also be steadfast in our willingness to set a performance benchmark that makes sense to us on meta-economic grounds—our knowledge of natural systems, our knowledge of the historical aberration of brief duration known as venture capital and its inextricable interdependence with all manner of hockey-stick growth, our intuition as to the benefits of diversity, diversification, and alternate courses of action.

Is 5 to 8 percent a lot or a little? Compared to what? Is this "venture capital minus" or "philanthropy plus"? Who will invest?

Institutional investors have been and will continue to be slow adopters in terms of food investing. (Slow, it seems, is not good in all things.) However, trends in the consumer and SRI marketplace bode well for slow money.

Although we might expect a few early-adopter institutional funders, primarily foundations, to provide seed capital, the ultimate market for slow money is individual investors: Investors' Circle types (angel investors interested in sustainability), shoppers at Whole Foods, local co-ops and farmers markets, CSA members, Prius owners, parents of Waldorf school students, and, more broadly, investors in SRI funds—that is, a much larger number of smaller investors. Many social investors are frustrated by the diluted impact of screened mutual funds. Despite its complexities from a fiduciary point of view, from the point of view of an individual investor who appreciates the value of a healthy local food system, slow money offers a kind of "pure play," targeting investment tangibly and directly at root drivers of sustainability.

It may also be possible and necessary to utilize charitable donations from individuals as part of slow money finance. We must be agnostic, at this early stage of design, as to sources of capital, types of securities, strips of financing. We have yet to fully affirm and define, with requisite particularity, our target. We have, ahead of us, some rather substantial feats of financial engineering and fiduciary entrepreneurship.

From all the meta-economics and macro-portfolio strategizing, we come down to deal flow.

The following companies, which have been in the market for financing in recent years, are broadly illustrative of the *type* of small food enterprises that might be candidates for slow money.

- AgraQuest is a biotechnology company that focuses on discovering, developing, manufacturing, and marketing environmentally friendly, natural pest-management products for the agricultural, institutional, and home markets. AgraQuest developed and commercialized its line of pest management products after screening more than twenty thousand microorganisms.
- Bion Waste Management provides an environmentally benign solution to air and water emissions from livestock-farming operations, transforming waste products into organic fertilizers and animal feed.
- Bioshelters and Waterfield Farms International are integrated production and marketing businesses, growing organic, sustainably produced herbs (mainly organic basil), fish (tilapia), and freshwater shrimp in linked aquaculture and hydroponics systems.
- The Farmers Diner seeks to revitalize local agricultural economies through building a chain of diners that serve foods sourced from within a 100-mile radius of each diner.
- ForesTrade is a supplier of organic tropical spices, essential oils, vanilla, and fair-trade coffee.

- High Desert Foods, which operates a 42-acre organic farm on which it farms nearly 9,000 fruit trees and 90 seasonal vegetables, is developing a line of organic pasta sauces, organic juice concentrates, and snack nut mixes.
- Honest Tea is building a brand of low-sugar, organic bottled teas, sourced in collaboration with indigenous peoples.
- Indigenous Designs has been a pioneer in fair-trade production and wholesale of handcrafted natural-fiber clothing, working through fair-trade alliances with artisan cooperatives throughout South America and India.
- Lightfull Foods offers healthy, natural snacks developed around its proprietary platform, the "Science of Satiety."
- New Day Farms is integrating an innovative, energy-efficient greenhouse-system with state-of-the-art soil enhancements, to produce organic heirloom tomatoes.
- Niman Ranch is a national brand for gourmet beef, pork, and lamb—meat that comes from animals raised by a network of family farmers who treat their animals humanely.
- Numi Tea is a rapidly growing purveyor of organic, artisan and flowering teas.
- Once Again Nut Butters processes peanuts, almonds, cashews, and sesame seeds into nut butters and roasted nuts, and packages honey.
- Pura Vida Coffee is a specialty coffee company that sells 100 percent certified fair-trade, organic coffee to individual and institutional consumers in the United States

in order to provide direct help to at-risk children and families living in coffee-growing countries.

- Wholesome Harvest, a coalition of small farmers, offers premium organic-certified poultry and meats to grocers, chefs, and households.

Back down to earth, immersed in deal flow, we can appreciate the energy of the entrepreneurs who are forging the relationships of the emerging restorative economy. Here, we sense the possibilities for a new kind of investing, a new kind of investor.

In Puebla, Mexico, this past November, at Slow Food's Fifth International Congress, Carlo Petrini spoke, playfully, of *engundia*, or sacred passion. Let us dare to imagine an investor who has the sacred passion of an earthworm, slowly making his or her way through the soil of commerce and culture, playing a small, vital role in the maintenance of fertility.

Now, whether such notions have any practical import to the task of creating this new entity called Slow Money seems, at first, implausible. But it isn't so. The success of Slow Money will depend on a vision that dares to be playful; that dares to assert a connection between *human* and *humus* and *humility* and *humor*; that dares to push back against the dismal science of economics and the hegemony of market mind; that dares to put money in its place, that calms money, in much the same way that "traffic calming" is becoming part of the agenda for "smart growth" in progressive communities, so that healthy relationships can once again begin to flourish; that dares to put "taste" and "flavor" back into investing, moving from the Big Mac school of fatty and salty buy-low/

sell-high investing to the Coleman Carrot school of investing that celebrates the subtle joys of *terroir* and authenticity.

The investor as earthworm versus the investor as Master of the Universe.

For my part, I would rather emulate an earthworm than an astronaut. I would rather enrich a plot of land than grow the portfolio of a quant.

Does this make me a bad investor? No, it makes me a different *kind* of investor. I would argue that an investor whose financial activity is predicated on extraction—on the linear take-make-waste methodologies of a world that had never seen the picture of the earth rising over the moon—is not really an investor at all. He is not truly investing himself in that to which he is applying his capital. Quite the opposite. He is keeping himself completely out of it, denying any personal connection to or responsibility for that to which his capital is lending its energy. How is it that something that is all about exits and liquidity and anonymity can really be called, in good faith, *investing* at all?

As I have written this essay, and up until the preceding few paragraphs, I have abjured or edited out the first person in many places—"I" this or "I" that—in service of a more professional, a more objective voice. Ironic, that as much as I understand that slow money is about reconnecting producer and consumer, investor and community and land, I am, still, so slow to realize that the connecting and convening functions of Slow Money may be as important as its fund-management strategies. People and place first, money second. Connecting producer and consumer and investor, not only in the abstract, but in person.

We are trained to take ourselves out of the equation, which is only the inevitable conclusion of an industrial logic that trains us to take ourselves out of our natural context, away from any one piece of land, away from any one place. This is surely why Schumacher chose "Economics as if People Mattered" as his subtitle.

It is as if, in the shadow of markets and money, we barely exist. We have become smaller, even, than earthworms. Smaller than mycorrhizae. We have become, Invisible Hand and all, invisible.

Somewhere on our way from Hiroshima to *An Inconvenient Truth*, some of us began to sense that the Invisible Hand no longer had clothes. It began to dawn on us that in a world that is heating up and speeding up, the health of future generations is no longer synonymous with the efficiency of capital markets, or, even, with the power of technology.

Seeing the Invisible Hand without its veils of protective ideology is as unnerving as seeing a naked emperor or a naked arbitrageur. We are going to need spiritual and emotional sustenance on this journey.

If the idea of slow money is going to take root, it is going to be in part due to inspiration derived from the celebratory, life-affirming, pleasure-inducing humanism of Slow Food. Slow Food began as a protest against McDonald's, but it quickly evolved from a protest into an affirmation, a series of creative initiatives dedicated to restoring and preserving quality of life. Similarly, Slow Money seeks to support the creative power of entrepreneurship to build new commercial relationships that enhance quality of life for farmers, food consumers, and their communities. In a world of special interests, policy debates, subsidies, and regulatory nightmares, the emergence of for-profit social entrepreneurs, who seek to build independent companies that integrate private enterprise and public benefit, is particularly intriguing and worthy of support. And part of what is so wonderful about many of these small enterprises is, to borrow the words of Ben Cohen, that they are places where people "can bring their hearts and souls with them to work."

Just as is the case with Slow Food, our cause is cultural, agricultural, economic, historical, biological, and even, perhaps, spiritual. Whether speed enables specialization or the other way around, who can say? But speed and technology and industrial compartmentalization are part and parcel of the same system of production that has carved today's capital-hungry society into a wobbly amalgamation of economic interest groups—employee, investor, shareholder, beneficiary, consumer—among which exist attenuated relationships that are "special" in only a perverted sense of the word.

If slow money is to become a useful tool in the toolkit of the sustainability-minded investor, it will need to be used with an appreciation of these "qualitative distinctions" and with a sense of urgency that reflects the global situation and millions of troubling local situations. Every investment we make is a statement of intention, a statement of purpose, a speculation about the future of man and his role in the scheme of things, not merely a financial speculation.

Every investment we make is a small part of larger questions:

- *What comes after industrial finance?*
- *Is there a healthier, more integral, more effective alternative to Wealth Now/Philanthropy Later?*
- *What if you had to invest 50 percent of your assets within 50 miles of where you lived?*
- *If it is prudent to invest tens of billions of dollars a year into a few thousand high-tech companies with explosive growth potential, then mustn't it also be prudent to invest a few billion dollars a year in tens of thousands of small food enterprises that are essential to the long-term stability and health of soil and community?*

PART TWO

ground zero

FOUR

in the shadow of the twin towers

"The market demands growth."

The speaker was Adam Werbach, former president of the Sierra Club, on the dais of San Francisco's Commonwealth Club, in response to an audience question about the limits of consumer activism.

It was February 1, 2007, a little over two years since he had made considerable waves at the same venue, delivering a speech entitled, "Is Environmentalism Dead?" Now, he is consulting to Wal-Mart as the commercial behemoth articulates its own version of an environmental imperative.

"The market demands growth."

A hundred times, a thousand times—how many times had I heard this axiomatic statement or one of its myriad corollaries? Axiom? Corollary? No: mantra. If free-market capitalism can *have* mantras, this is surely one.

"The market demands growth."

I cannot say just what it was that made this particular encounter with these words so powerful.

No, there was no sudden, traumatic impact of poet hijacker slamming into towers of economic prose and fiduciary power. Neither was there any shudder of epiphany. It was as if a final veil, a stubborn little, final veil, had finally fallen from the icon of capitalism, around whose feet I had been circumambulating, invisibly, quixotically, stoically, with self-doubt-studded irreligious persistence for thirty or forty years now. On the words of this somewhat brash young "post-environmentalism" consultant-cum-activist, it fell, it fluttered to the floor, this last little veil, and with it went the last of my inhibitions, my left-brain bugaboos, my taboo deference, my doubts. It was as if I had had the presence of mind to walk up to the dais after the talk, and, while handfuls of attendees jammed around the speakers for this reason and that, I had stepped to the side, stooped down, picked it up and stuffed it into the pocket of my Patagonia fleece.

And here I am on this fine, northern New Mexican morning, a few weeks later, the veil before me on my desk, awaiting illumination by the sun as it rises behind Truchas Peak, spilling pale peach, yellow, orange, blue, and gray before it and hanging a purplish haze, as if it were mist, above the stone gray backdrop of the Jemez Mountains and the rim of the Valle Caldera forty miles to the west.

It was on Super Bowl Sunday that I returned from my trip to San Francisco. I stopped at La Montanita Co-op in Santa Fe to stock up on food on my way home, and it was already 4:30 when, driving through Nambe, I got a call from my son in Philadelphia.

"You watching it, Dad?"

"Not home yet. Will catch the second half."

I was a little behind schedule, and hurrying to get home while the passive warmth of the day was still held by glass and logs and stucco and adobe, so that my work with the woodstove would have a little less urgency about it.

A few days earlier, one of my neighbors had said to me on the phone, "Yep. We had another ten inches or so. With lots of drifting. It's supposed to go down to eighteen below zero tonight."

I was pleased to see that there was no evidence of cow incursion.

My acre of land is situated amid hundreds of acres of pastures across which small herds range without a hint of urgency. Theirs is an almost palpable sense of entitlement that transcends impunity. Although land ownership is demarcated by barbed wire, land use seems, from my year or so of casual observation, to be informally and relatively communally managed. The herd that is usually nearby is cared for by Ben Sandoval, who drives up from Chimayo most days during the winter to make sure water and hay are ample.

Cattle move from pasture to pasture, sometimes, it seems, perhaps most times, of their own accord. It is not unusual to find several head or an entire herd ambling down County Road 79, poking their heads under barbed wire fencing after a few blades of grass that may have been revealed by receding snow. It is not unusual to find a few head of cattle in my yard, testing the fencing around my future vegetable garden, munching on the tiny branches of my new fruit trees or leaving randomly placed contributions to soil fertility where they may or may not be appreciated as organic reminders of their considerable presence.

The house was cold, but the faucets ran fine, the cats were happy, and there was enough sunlight left to suggest that by the time nightfall arrived, the worst of the chill would be gone.

Splitting firewood, carrying firewood, starting the fire, tending the fire: In the six years that I have been living off the grid, utilizing a woodstove as my sole source of heat (save for a propane-fired space heater that keeps the house from freezing when I am away), I have never considered these chores onerous or inconvenient. There is something pleasing about them, a ritual, manual, down-to-earth rhythm that punctuates the day as beautifully as any semicolon ever graced a compound sentence.

Compare these two acts: adjusting the thermostat versus tending the fire. Even the words *adjusting* and *tending* point to deeper meanings. Adjusting in this context is a mechanical term, tending more a term of care.

Almost without fail, my interaction with firewood puts me in a positive frame mind, as if (bringing in another meaning of the word *adjust*) it were adjusting, or adapting, my relationship to the rhythms of natural systems. Adjusting a thermostat has never had any such salutary affects on my psyche.

The very sight of a woodpile is pleasing to me, and it was no different on Super Bowl Sunday as I opened the gate, pulled in from County Road 79, and drove the hundred yards from the dirt road to the stone path that leads to my front door. At the far end of the property, in clear view, is a two-cord pile of split wood, which will in all likelihood not be touched until next winter. The remainder of this year's supply is at the side of the house, protected by a shed roof from the elements.

I've often wondered how much standing timber makes a cord. How many cords to a mature tree? How many trees, or what part of a mature tree, to a winter's warmth? In cords, for this particular domicile, I have a rough answer: two.

It is not important to me that I know the precise answer to such questions, however; the quantitative answer is not necessary for, does not inform my appreciation. I do not need to *know* in order to *appreciate*. In fact, a mind cluttered by information posing as knowledge—indices of consumption and torrents of disaggregated bytes that become "data smog"—too often gets in the way of appreciation. Knowledge is the mind-set of adjustment and consumption; appreciation is the mind-set of tending and care. Appreciation is rooted in a sense of doing as little harm as possible, of deriving satisfaction from small pleasures, pleasures that have little or nothing to do with the marketplace.

And so, thoughts and feelings can flow while returning to a rural household, entering into the presence of firewood and fire, woodstove and chimney.

And so, my thoughts and feelings flowed as I pulled up to the house, appreciating my return from the city and the realm of professional environmentalists and post-environmentalists, industrialists and post-industrialists, urban consumers and would-be consumer activists, Wal-Mart changers and Wal-Mart bashers, champions of what Peter Barnes calls *Capitalism 3.0*, members of what one observer has called the "cultural creative" psycho-demographic, attendees of green-business conferences and social-entrepreneurship conferences and the like. I immediately set to making a fire, but this time there was something else on my mind.

The echo, the shadow of Adam Werbach's words: "The market demands growth."

My trip had, in fact, encompassed another meeting.

Prior to going to San Francisco, I had participated in a retreat in the Utah mountains, where approximately two dozen entrepreneurs, investors, and philanthropists had been come together to explore a wide range of issues.

"So many relationships have been broken. I believe that media can play an important role in healing them."

The speaker, an entrepreneur of Arabic descent, described how and why he is dedicating substantial amounts of time and wealth to producing films about the clash of civilizations between the West and Islam.

What is 9/11 about but broken relationships? What is global warming about but the broken relationship between *Homo sapiens* and the biosphere? What is Buy Low/Sell High about but the ultimate of relationship breakers: a financial calculus that reduces everything to a commodity and a transaction? What is Internal Rate of Return about but the professional investor's horribly powerful, hegemonic razor, severing long-term relationships between investor and company, and between company and community?

The American Way of Life, which the first President Bush called, in defiance of the Kyoto treaty on climate change, "not negotiable," is built around the Holy Grail of unlimited economic growth. There is no such thing as too much economic growth. There is no such thing as too much consumer confidence. There is no such thing as too much wealth. There is no such thing as a city that is too big. There is no such thing as a company that is too big. Quality of life is synonymous with maximum growth and maximum consumption. Broken homes, broken cities, broken accounting rules, broken Main Streets in rural communities across the land, broken endocrine systems, broken watersheds: Broken relationships do not, must not, in the name of the health of The Economy, preempt the growth of The Market.

Suddenly, I see the virtual twin towers of the economic worldview.

"The market demands growth." This is one virtual twin tower.

"The American way of life is not negotiable." This is the other virtual twin tower.

In the shadow of ground zero, unshaken, defended by preemptive military might and a hundred thousand lives and trillions of dollars, they stand, tall as ever.

Despite significant growth and dynamism in the fields of social investing, philanthropy, and venture capital over the past few decades, we have done little to escape the shadow of these twin towers. We continue to grow classes of professional deal doers and classes of professional do-gooders, but the partial views and financial imperatives of their professional disciplines prevent them, and their organizations, from responding wholly to the situation that is unfolding so very rapidly around the world. "Front of the pipe" investing pursues explosive growth at a more frenzied pace than ever, while "end of the pipe" philanthropy remains deployed in a rear-guard action, putting band-aids on the deep wounds caused by globalization and consumerism and the ever more overheated markets that serve them.

Modern philanthropy is dogged by a dissonance that arises from an inherent contradiction: Philanthropic assets are invested without concern for a foundation's mission, with only the earnings, or

5 percent, used to fund grants. This 95/5 split and the cultural iron curtain between those who give grants and the money managers who oversee assets highlights starkly the stubborn disconnect between private enterprise and public benefit.

A similar "brokenness" dogs the entire social-investment field. Even the most progressive social-investment funds feel compelled to seek competitive returns. With a few exceptions (such as Highwater Capital, a newcomer inspired by the research of Paul Hawken and managed by Baldwin Brothers Investments), these funds find themselves with portfolios that include the likes of Wal-Mart or McDonald's or an oil company that has women and minorities on its board of directors. Virtually the entire social-investment field has signed on to the motto Doing Well While Doing Good, meaning that the prudent social investor can achieve competitive financial returns while avoiding such things as sin stocks, bad corporate apples, or egregious polluters.

Thus, philanthropy and social investing remain stuck in the shadow of "having your cake and eating it to."

"But don't knock the idea of 'doing well by doing good,'" says John Fullerton, CEO of the hedge fund Alerian Capital and formerly of JPMorgan Chase. "While I agree that it is an inadequate response to much more profound problems, at least it is an incremental response. It is part of a common attempt to express the psychosis that we all feel deep in our guts."

The terrible truth—and this has been well documented by such groups as Redefining Progress—is that economic growth brings along with it, in addition to wealth, all kinds of what John Ruskin long ago called "illth." Rising divorce rates, homelessness, brownfield remediation, urban blight and suburban sprawl, smog, smoking-related illness, the dead zone in the Gulf of Mexico, an epidemic of obesity in the United States, child labor in developing countries, and one new coal-fired power plant a week in China: These are just a few of the "side effects" of an economic system that started, harmlessly enough, by bringing us two chickens in every pot, graduated to two cars in every garage, and is now shooting us into the stratosphere of a Starbucks on every corner and unprecedented stock market heights.

To many of us, it appears increasingly obvious that as we head toward seven billion global inhabitants and past four hundred parts per million of carbon in the atmosphere, the ratio of illth to wealth is becoming perilously skewed. Or, put another way, the considerable benefits of global capitalism are coming increasingly at life-threatening social and environmental costs.

"The market demands growth."

Yes, it is true.

Capital markets as we currently know them arose in a world of five hundred million people and vast frontiers, a world in which commerce was small and natural resources seemed infinite, a world out of which the industrial revolution was about to explode and the idea of a "limited liability company" was an unprecedented innovation, marvelously successful at unleashing floods of investment capital in order to support the once-in-the-history-of-the-planet growth that would occur in the ensuing two centuries. What was in the 1700s and 1800s a tremendous innovation, the limited liability company, has morphed into the modern, multinational corporation and armies of professional fiduciaries—a financial system of such scope, size, complexity, and impersonal power that it could never have been envisioned by the framers of the Constitution, in which the word "corporation" never appears.

It is not surprising that we find ourselves, today, captive to capital markets that are themselves captive to the enormous momentum of the economic growth that they have made possible.

What is surprising, however, is the degree of our reticence, our impotence, our unwillingness—in the face of the collision course between unlimited economic growth and the limits of culture and the biosphere to absorb our accelerating levels of extraction, consumption, and pollution—to dare to imagine another way.

"Sometimes I think we are trying to change the direction of the wind," said Bart Holaday, an Investors' Circle board member. An interesting blend of hard head and soft heart, this former Air Force officer, this retired oil and gas investor and former manager of billions of dollars of institutional venture capital now spends time putting an electric engine in an old pickup truck, pursuing a philanthropic agenda in his home state of North Dakota, and putting money into double-bottom-line venture deals.

"Riding the wind in a slightly different direction is one thing," Bart said. "Trying to change the direction of the wind? Now, that's something else entirely. It can't be done."

Yet that is precisely what we must dare to do. . . .

Well, not exactly.

Because unlike the wind, financial markets are man-made.

And we can remake them.

So, I am sitting here, contemplating the veil that lies on my desk. It seems harmless enough. Yet what it prevents us from seeing is not. Behind this veil, and behind the iconic devotion to the market that it conceals, is . . . is . . . well, if I thought I could get away with it, without seeming to trivialize this very important subject, I'd flash to Stephen King's *The Langoliers.*

Instead, I'll land somewhat closer to home with a few essayist-poets. Because the wisdom we need to reconnoiter, to find our true bearings, to see things whole, must come from outside of economics, from above and beyond the market.

First, Gary Snyder:

> We also see that we must try to live without causing unnecessary harm, not just to fellow humans but to all beings. We must try not to be stingy, or to exploit others. There will be enough pain in the world as it is.
>
> Such are the lessons of the wild. The school where these lessons can be learned, the realms of caribou and elk, elephant and rhinoceros, orca and walrus, are shrinking day by day. Creatures who have traveled with us through the ages are now apparently doomed, as their habitat—and the old, old habitat of humans—falls before the slow-motion explosion of expanding world economies. If the lad or lass is among us who knows where the secret heart of this Growth Monster is hidden, let them please tell us where to shoot the arrow that will slow it down.[1]

Second, Wendell Berry:

> The corporate mind is remarkably narrow. It claims to utilize only empirical knowledge—the preferred term is "sound science," reducible ultimately to the "bottom

1. Gary Snyder, "The Etiquette of Freedom," in *Practice of the Wild* (New York: North Point Press, 1990).

> line" of profit or power—and because this rules out any explicit recourse to experience or tradition or any kind of inward knowledge such as conscience, this mind is readily susceptible to every kind of ignorance and is perhaps naturally predisposed to counterfeit knowledge. It comes to its work equipped with factual knowledge and perhaps also with knowledge skillfully counterfeited, but without recourse to any of those knowledges that enable us to deal appropriately with mystery or with human limits. It has no humbling knowledge. The corporate mind is arrogantly ignorant by definition.[2]

What we are facing is not an economic disease, but an economic symptom of a cultural disease. Our problem is not an economic problem, but a problem of how we think and how we live. The solution lies not in the hands of economists and investment bankers, but in the hearts and minds of poets and in the portfolios of every man and woman who puts money into "the market."

Berry points to what may be the single most important missing ingredient in the modern, economic-growth-obsessed mind-set: humility. Is there such a thing as a humble corporation?

Now that would be a company in which I would like to invest.

Where can I go to find such an investment?

Surely not to Wall Street.

Surely not to the market that generates a $53-million bonus for the CEO of Goldman Sachs and a $400-million severance pack-

2. Wendell Berry, *The Way of Ignorance*, (Berkeley, CA: Shoemaker & Hoard, 2005).

age for the outgoing Exxon CEO. Surely not to the market that holds hundreds of billions of dollars for foundations that cannot see how to deploy these assets in fundamentally new ways and trillions of dollars for pension funds that are prohibited by law from considering anything other than traditional financial factors in their investment decision making. Surely not to the marketplace inhabited by legions of brokers, bankers, traders, arbitrageurs, hedge-fund managers, sector analysts, day traders, specialists, short sellers, margin callers, and investment advisors whose dedicated aim—protected wholly and unconditionally by the veil of Adam Smith's Invisible Hand—is to maximize their own compensation by maximizing financial throughput.

This is not a market that is fundamentally aligned with my values, and in dealing with it, I always feel fundamentally uncomfortable, fundamentally compromised. As long as unlimited economic growth is Mantra One, no amount of screening can create a portfolio that overcomes my *dis*-ease.

This is not, as it would be all too easy to aver, because most folks are greedy. "Greed" is a gross oversimplification of forces that are far more pervasive and far less easily categorizable as evil. Is the $225 billion California Public Employees' Retirement System greedy when it tries to grow assets under management? Even if there are those among us who are troubled by the implications of pools of capital of this scale, one would be hard-pressed to attribute its continued growth to the greed of individual pension-fund trustees or individual pension-fund beneficiaries.

No, I don't think one can attribute the orientation of the market to the greed of most individuals. More, it is a result of fear, uncertainty, and insecurity. In a world in which producer has been divorced from consumer, in which most of us produce virtually nothing or actually nothing that we consume, an ever more complex and impersonal and fast-paced world, a world in which extended webs of social relationships are being increasingly attenuated in favor of virtual relationships, we have little to fall back on for our immediate, day-to-day material security but purchasing power.

Relationships are being rent asunder, relationships are under siege, and the market offers as compensation the only thing it can offer, a lowest-common-denominator, one-size-fits-all, to-a-hammer-everything-looks-like-a-nail solution: economic growth.

What is to be done?

The answer is as simple as it is complicated.

We need to create a new market. A new *kind* of market. A market that will turn *the* market into *a* market or *that* market.

We need a market that rewards humility and promotes patience and invites the participation of all those individuals who will sleep better at night knowing that some of their dollars are swirling around cyberspace a little bit slower, lending a little bit less of their energy to the economic engine that brought us, last year, 8 million light trucks and SUVs and who knows how many million Twinkies. We need a peaceful market, a market that rewards peaceful companies, a market that dares to recognize explicitly, publicly, and financially, that growth, growth, growth is predicated on dislocation and churn and continuously reinvented and unsatisfiable consumer demand, and that these conditions constitute a form of economic violence.

We need a market that will be to Wall Street what Switzerland and Costa Rica are to the Pentagon.

We need a market that rewards companies that do not build

value on broken relationships and whose value-creation process is built around the preservation and restoration of relationships: relationships between individuals, relationships between producers and consumers and communities, relationships between cultures, relationships between species. We need a market that puts the *share* back into shareholder, recognizing that it is no longer advisable, no longer prudent, or, even, no longer moral for the current generation of the beneficiaries of economic growth to take all or most of their profits off the table for their own use, relegating the solutions to the great challenges of our time to the realms of politics and philanthropy.

Let us drop down from such dreamy rhetoric and see if we can begin to imagine what some of the design elements of such a market might be.

I am imagining what we shall call, for the sake of this thought experiment, the Main Street Exchange.

As I embark on this imagining, I do so fully cognizant of the extent of my ignorance. My few decades of experience as an entrepreneur, small-time venture capitalist, foundation treasurer, and activist has had me in and around some of our nation's most sophisticated and successful investors, but my learning and technical financial capabilities have many gaps in them.

I am going, therefore, to invoke a veil to cloak my ignorance. A very different kind of veil than the one that fluttered down a few weeks ago at the Commonwealth Club in San Francisco. This would be a Thoreauvian veil, a veil behind which ignorance becomes, mysteriously, reconnected to wisdom.

Thoreau wrote, "How can he remember well his ignorance,

which his growth requires, who has so often to use his knowledge?" Harkening back to Wendell Berry's admonition about the narrowness of corporate knowledge, I think, as I echo Thoreau's words, of today's ultrasophisticated fiduciary, who has so often—almost incessantly, as capital moves faster and faster—to use his knowledge.

I imagine that the Main Street Exchange would be organized around such principles and operating guidelines as the following:

1. It is time to reimagine business as a tool, first and foremost, for rebuilding communities and restoring bioregions. The Main Street Exchange will be the first stock exchange dedicated to building the *restorative economy*.
2. The global economy would be far more sustainable, communities more vital, and cultures richer and more diverse, if tens of thousands of small, independent, locally rooted companies had greater access to capital and millions of small investors had the ability to invest in them. The Main Street Exchange will design its investment products and services to serve this market.
3. The primary purpose of the exchange is not to promote maximum economic growth, but rather to create liquidity for investors and access to capital for enterprises that accelerate the transition to a restorative economy.
4. The exchange will develop requirements for listing eligibility with respect to a company's stakeholder accountability, corporate governance, charitable giving, and other aspects of a clearly stated commitment to sustainable business practices.

5. The exchange will explore structures and incentives that reward investors for low turnover and long holding periods.
6. The exchange will develop a Patient Capital Index.
7. Fifty percent of the profits of the Main Street Exchange will be donated to a nonprofit MSE trust, which will be utilized to support the development of *local stock exchanges* and other innovations in intermediation that support restorative economics.
8. The exchange will encourage all listed companies to donate 50 percent of their profits to similar trusts, to be deployed as decided by company management, employees, and directors.
9. Shares sold through the Main Street Exchange will have no voting rights. Voting rights offer individual investors the verisimilitude of voting control, but de facto voting control is exercised by a small number of institutional fiduciaries, which control is at odds with the *independence* of listed companies.
10. The exchange will designate target percentages of listed companies with respect to such sectors as organics, local food enterprises, regional renewable-energy suppliers, community-development ventures, and other enterprises that are vital to community and bioregional health, but that are traditionally considered too small for public ownership through a national exchange.
11. There *is* such a thing as a company that is too big. Beyond a certain scale, companies lose allegiance to any particular place and become beholden not to the needs of people, but to the demands of unfettered capital.

> The Main Street Exchange will seek to develop principles of *appropriate corporate scale* that can be applied to companies as part of maintaining their eligibility for listing.

There is an increasingly robust wave of entrepreneurial activity around such principles and concerns. Hundreds of mission-driven early-stage companies every year seek capital through the Investors' Circle (www.investorscircle.net). Scores of cities are organizing chapters of the Business Alliance for Local, Living Economies (www.balle.org); BALLE is incubating the concept of local stock exchanges. B Lab is incubating legal guidelines for B corporations and formulating an entrepreneurial strategy for accelerating the growth of a B sector (www.bcorporation.net). Inspired by Newman's Own, a handful of intrepid, celebrity-less companies, including Untours, Pura Vida, and Give Something Back, dedicate all or a substantial portion of their profits to charity. The Fourth Sector Network is exploring the development of "for-benefit" organizational forms and governance structures.

Absent the liquidity and established reporting guidelines of an organized stock exchange that is dedicated to this marketplace, these companies and their investors are plagued by all kinds of ambiguity. But they are a harbinger of what is to come after the last veil of capitalism has fallen (and the Thoreauvian veil of ignorance has been picked up).

Is it really possible to design a market that does *not* demand growth?

We must dare to ask the question and we must dare to answer it.

This is not a problem that contravenes the laws of physics or biology. This is a human institution. This is a problem of design. This is a challenge to hold our heads above the rising tide of industrial finance. With all the knowledge and ignorance at our disposal, just why in the world would we not be able to design an alternative market, no matter how complicated?

And if such a market were to come into existence, with however few listed companies and however few investors, to what scale might it appropriately grow over time?

Given the severity and complexity of the current global situation, the idea of a Main Street Exchange may seem fanciful.

Our way of life is under increasing threat. We are being directly and indirectly attacked. We need to protect and defend ourselves.

As a nation, we are looking to defeat our enemies and remedy our ills through the deployment of military power and the pursuit of economic growth. Yet when we defer to this process as individuals, we surrender, unwittingly, our power to implement fundamental social change, to go beyond alleviating symptoms, and to address root causes.

"Our complex global economy is built," observed Theodore Roszak, "upon millions of small, private acts of psychological surrender." Every dollar that we send into "the market," disconnected from our beliefs and our values, disconnecting us from one another, from our communities, from the land, is an act of surrender. No market victory, no Dow Jones Industrial Average record, can compensate for this surrender.

Perhaps this is one of the many things that I love about tending the woodstove. It is a small but potent therapeutic form of reconnection. It brings you close to the decisions that affect your life. It makes you pay attention. It facilitates surrender, but surrender of a different sort.

In "The Gift Outright," heard by all who witnessed JFK's inauguration in 1960, Robert Frost describes a form of surrender that holds salvation for the American people:

> Something we were withholding made us weak
> Until we found out that it was ourselves
> We were withholding from our land of living,
> And forthwith found salvation in surrender.

Living in a place, as I am fortunate enough to do, where forests are plentiful and humans are relatively few and far between, the woodstove and the woodpile foster a gentle, practical, household-healing form of surrender.

Heating oil in an oil tank in a basement? You never see it, smell it, touch it. It never sees the light of day. It comes from far away, from places that are invisible to you, at costs political, economic, and environmental that are both calculable and incalculable.

Who knows how much oil there is still in the ground? Experts and corporations and armies will fight over it.

Meanwhile, I can see the trees that stretch from here through the foothills and up to the Pecos Wilderness. Seeing them, being close to them, and knowing that I depend on them, I must take care that my use of them, and their use by others, does not exceed their capacity to replenish themselves.

"The market demands growth."

I continue to contemplate these words. And the veil they caused to flutter down.

Can I put it to any real use?

Dust rag for my woodstove? Biodiesel rag for the converted pickup truck I don't yet own? Tiny banner to be emblazoned with the logo of the Main Street Exchange?

FIVE

the war on *terroir*

On this fifth anniversary of 9/11, the president proclaims that the events of that day have caused Americans to "see the world in a new way." The radio broadcasts stories of patriotism:

> I was just a boy in the 1960s. My adolescence wasn't infused with the civil rights struggle or the sexual revolution or the Vietnam War, but with their aftermath.
>
> My high school teachers were ex-hippies and Vietnam vets. People who protested the war and people who served as soldiers. I was taught more about John Lennon than I was about Thomas Jefferson.
>
> Both of my parents were World War II veterans. FDR-era patriots. And I was exactly the age to rebel against them.
>
> It all fit together rather neatly. I could never stomach the flower-child twaddle of the '60s crowd and I was ready to believe that our flag was just an old piece of cloth and that patriotism was just some quaint relic, best left behind us.
>
> It was all about the ideas. I schooled myself in the writings of Madison and Franklin and Adams and Jefferson. I

came to love those noble, indestructible ideas. They were ideas, to my young mind, of rebellion and independence, not of idolatry.

But not that piece of old cloth. To me, that stood for unthinking patriotism. It meant about as much to me as that insipid peace sign that was everywhere I looked: just another symbol of a generation's sentimentality, of its narcissistic worship of its own past glories.

Then came that sunny September morning when airplanes crashed into towers a very few miles from my home, and thousands of my neighbors were ruthlessly incinerated—reduced to ash. Now, I draw and write comic books. One thing my job involves is making up bad guys. Imagining human villainy in all its forms. Now the real thing had shown up. The real thing murdered my neighbors. In my city. In my country. Breathing in that awful, chalky crap that filled up the lungs of every New Yorker, then coughing it right out, not knowing what I was coughing up.

For the first time in my life, I know how it feels to face an existential menace. They want us to die. All of a sudden I realize what my parents were talking about all those years.

Patriotism, I now believe, isn't some sentimental, old conceit. It's self-preservation. I believe patriotism is central to a nation's survival. Ben Franklin said it: If we don't all hang together, we all hang separately. Just like you have to fight to protect your friends and family, and you count on them to watch your own back.

So you've got to do what you can to help your country

> survive. That's if you think your country is worth a damn. Warts and all.
>
> So I've gotten rather fond of that old piece of cloth. Now, when I look at it, I see something precious. I see something perishable.[1]

I wonder what is new in this way of seeing the world.

As a boy in the 1950s, my adolescence *was* infused with the civil rights struggle and the Vietnam War and the sexual revolution. The events of those years made it abundantly clear that the enemy was not a particular person or group, but violence in all its forms: military intervention in another country's affairs in the name of geopolitics; the futility and moral bankruptcy of trying to impose democracy by force; the vilification of political foes; fear of conspiracies; the horror and harm caused by individual zealots; the insidious violence to the social fabric done by political corruption.

These are not new issues. Only the incidents through which they have been brought home to us are new. Only a nation with the attention span of the e-generation could construe them as new. Only a nation with the attention span of a day trader could fail to see that the battles are new, but the wars are as old as agriculture.

For days, pictures of it were streaming through cyberspace and into our living rooms and kitchens and dens and bedrooms: the most

1. Frank Miller, "That Old Piece of Cloth," broadcast on National Public Radio's *Morning Edition*, September 11, 2006.

famous hole in the world—Saddam's spider hole. Symbolizing, to all, the demise of despotism, and, to some, the manifest destiny of a certain way of life.

Your garden-variety, run-of-the-mill gopher hole would seem, on the face of it, to have little to recommend itself by way of comparison: completely unnewsworthy, utterly lacking in geopolitical significance, and of zero entertainment value.

The gopher hole makes an unlikely symbol for the wars that will shape the future as we head toward a 7-billion-person planet and a $100-trillion-a-year global economy:

> *The Red War.* The battle against subsidy. The struggle to find ways to reduce the extent to which we are borrowing from the planet and stealing from future generations to fuel consumerism and economic growth. Think: red ink.
>
> *The Blue War.* The battle against intermediation and consumerism run amok. The struggle to reassert the primacy of the real over the virtual. Think: blue TV-screen light.

Living in an age that consumes, every day, fossil fuel that was 10,000 years in the making, can we think of ourselves as anything other than "in the red"?

Living in a country in which the average person spends four hours a day in front of the television being bombarded by commercial messages, distractions, and seductions, can we think of ourselves as anything other than "in the blue"?

The Red War and the Blue War have nothing and everything to do with the War on Terror. We can change regimes, capture the members of terrorist organizations, and keep hair gel off of every plane on the planet, all the while losing the wars that will determine our fate in much more fundamental ways.

The real enemy is not Communist or Muslim or media corporation or Republican or Democrat or Jew or Christian fundamentalist. The real enemy is a worldview that puts us deeper and deeper in the red and in the blue, a worldview that hides the challenges of the next century behind the glitter of the last.

The War on Terror absorbs our attention. Meanwhile, our fate is being determined by deeper struggles for our hearts and minds: allegiance to flag versus allegiance to land; allegiance to nation versus allegiance to place; allegiance to homeland versus allegiance to household.

This is the War on *Terroir*.

The French term *terroir* can be thought of as the ultimate expression of concern for a particular place, a particular piece of land. The word's literal translation is "soil," but it has come, courtesy primarily of the wine industry, to encompass much more. On one level, it refers to the soil qualities of a particular vineyard, or even a particular parcel within a vineyard—soil type, mineral content of the soil, pH level, drainage, macroclimate, microclimate, sun exposure, and so on. On another level, it has come to represent a range of environmental and cultural qualities that are specific to a place but difficult or impossible to quantify.

The subject is well encapsulated on the home page of the *Terroiriste*:

> *Terroir* is a subject of some controversy in the wine world. Virtually everyone would agree that climate in a macro

> sense influences the character and quality of wines. However, the significance of soil and microclimate is much debated. At one extreme are technologists, often but not exclusively from "New World" regions like the US or Australia, who contend that what really matters is varietal type, clonal selection, and wine-making technique, with technology and modern techniques available to manipulate the final result. A more moderate view admits the importance of vineyard management, especially control of yields, but expresses doubt concerning the overall significance of variations in precise soil type or microclimate.
>
> At the other extreme can be found the practitioners of biodynamics ("la biodynamie"), a quasi-mystical approach to winemaking that views the vineyard as a holistic living system and borrows from astrological methods to chart the "energies" of individual plots. Though easy to dismiss as spiritualist hokum, it is interesting to note that some of the finest wines in the world are produced by biodynamic partisans. If results are the true test of a system, biodynamics is deserving of serious attention.

Taken broadly, the concept of *terroir* is about much more than the quality of wine from a particular region. *Terroir* challenges us to become more nuanced in our appreciation of, and, therefore, our desire to preserve, the qualities of particular places, particular ways of producing food, and the local cultures to which they are integral. It asks us to look beyond "modern techniques" and "clonal selection" and the hegemony of brands, to rediscover a respect for relationships, a celebration of diversity, and an affirmation of the rightful place of humility and mystery alongside our impulse to control.

Wine, the elixir of the gods—where better to find our bearings in a world of fateful choices?

Which wine are we going to choose? The wine of technologists, militarists, fundamentalists, Islamists, consumerists? The wine of flag-waving *homelandistas*? The wine of terror and fear? Or a different kind of wine altogether? The wine of cultural and economic diversity, the wine of tolerance and forgiveness, the wine whose delicate bouquet holds intimations of the slower, the smaller, the less uniform, the less commercial, the less violent?

The events of 9/11 and the War on Terror are, most assuredly, the foreground of this moment in history; but in the background the War on *Terroir* acts out violence of a less newsworthy kind—the intentional and unintentional violence of a global economy that unsettles communities and degrades the land.

For a few centuries now this War on *Terroir* has been unfolding in the shadows of the Industrial Revolution, impelling people around the world, by the hundreds of millions and billions, to choose urban over rural, markets over community, consumer power over self-sufficiency, speed and mobility over permanence and stability, the mechanical over the biological, the technological over the natural, monoculture over diversity, virtual over real. The American Dream was a perfect accelerant for these processes.

A few hundred years ago, corporations were small, the world was big, and technology was in its infancy. Natural resources seemed vast and the idea of global ecological disruption was utterly inconceivable. It made sense, then, to optimize around the creation of markets, the efficient manufacture of goods, and technological innovation.

Today, corporations are huge, the world has become small, and technological innovation is exploding. Indicators of global ecological disruption and social instability are piling up. It makes sense to "see the world in a new way," to look beyond geopolitics and GDP, to optimize around the health of communities and bioregions and the preservation of the commons.

It is time to recognize that the ways of thinking, the beliefs, the allegiances that made sense during the Industrial Revolution no longer serve us well in this Age of Global Markets, this Nuclear Age, this Age of Ones and Zeros.

We have seen the earth rising over the moon: What do we wish to *do* with this information? We have learned that soil fertility depends on myriad biological relationships, not just on the levels of N-P-K that can be applied chemically: What do we wish to *do* with this information? We have learned that although global markets create a thousand billionaires, they do not necessarily improve the lots of a few billion thousandaires: What do we wish to *do* with this information?

Am I a *terroiriste*? Perched here in the New Mexico mountains, off the grid, yet connected electronically to a network of like-minded investors and philanthropists around the United States, am I a nonviolent guerilla fighter against the destructive extremes—extremes of militarism, extremes of religious divisiveness, extremes of consumerism at all costs—toward which the American Dream is swinging?

The metaphor of a pendulum is comforting, at first, but then this particular pendulum seems to have too much of the wrecking ball about it. By the time it stops and returns, many of us—us *Homo sapiens* and our descendants, us plants and animals, us worshipers and our gods, us co-inhabitants and stewards of this globe—will be robbed of the opportunity to enjoy what is left.

I remember Gary Snyder's entreaty in "The Etiquette of Freedom" to shoot an arrow at the heart of the Growth Monster.

An apparently quaint, notion, this—that an arrow aimed right could slow the Growth Monster. *A peashooter against a stealth bomber? No. The English versus the Spanish Armada. David and Goliath. Tiananmen Square.* Then my thoughts come much closer to home (but not quite as close as a gopher hole)—to the elk that inhabit, still, the Sangre de Cristo Mountains a few miles to my east.

The arrow is where ancient and postmodern meet, where hunter-gatherer and physicist meet:

> The concept of linear time carries with it the implication of an *arrow* of time, pointing from past to future and indicating the directionality of sequences of events. . . . Scientists and philosophers have been sharply divided over the significance of the arrow of time. The conundrum, put crudely, boils down to this: is the universe getting better or worse? The Bible tells the story of a world that starts in a state of perfection—the Garden of Eden—and degenerates as a result of man's sin. However, a basic component of Judaism, Christianity, and Islam is a message of hope, of belief in personal betterment and the eventual salvation of mankind. . . .

> The second law of thermodynamics introduces an arrow of time into the world because the rise of entropy seems to be an irreversible, "downhill" process. By an odd coincidence, just as the bad news about the dying universe was sinking in among physicists, Charles Darwin published his famous book *On the Origin of Species*. . . . Biological evolution also introduces an arrow of time into nature, but it points in the opposite direction of the second law of thermodynamics—evolution seems to be an "uphill" process. Life on Earth began in the form of primitive micro-organism; over time, it has advanced to produce a biosphere of staggering organizational complexity, with millions of intricately structured organisms superbly adapted to their ecological niches. Whereas thermodynamics predicts degeneration and chaos, biological processes tend to be progressive, producing order out of chaos. Here was optimistic time, popping up in science just as pessimistic time was about to sow its seeds of despair.[2]

These arrows and Gary Snyder's reference to the arrow that can kill the monster of economic growth make me wonder whether the economy itself isn't an arrow. To the arrow of time, and the conundrum of its heading toward both degeneration and decay on the one hand, and increasing complexity and organizational sophistication on the other, let's add the arrow of economics. The arrow of economics presents us with an analogous conundrum: *Is economic growth making the world better or worse?*

2. Paul Davies, *About Time: Einstein's Unfinished Revolution* (New York: Touchstone Books, 1995).

It is as tempting to scoff an immediate, "Well, of course, it is making the world better!" as it is to posture simplistically about smoking all the terrorists out of their holes.

But the truth, at this historic juncture, has become double-edged.

Is it true that China is building one new coal-fired power plant each week? That heart disease is exploding among the Chinese population because of fast food and smoking? Are we on our way to a one-billion-car, rainforest-less planet? Would the United States be socially and environmentally better off with no small farmers and the vast majority of its food imported from countries where it can be produced less expensively? Does continued media consolidation lead to higher quality, more diverse programming? Is the world better off if all the peasants leave the countryside for the cities to drink Coke and get jobs making shower curtains for export to the United States? Has the general sense of well-being in America gone up over the past few decades in tandem with the overall rise in GDP? Is it a welcome sign of progress that so many Main Streets in so many rural towns are littered with abandoned storefronts? Is it a sign of cultural health that there are more malls than high schools in this country?

Venture capitalist John Doerr's fateful phrase "the greatest legal accumulation of wealth in history" is an apt description, but an incomplete one.

We are also confronted with the greatest accumulation of poverty in history, the greatest accumulation of carbon in the atmosphere in history, the greatest accumulation of Hummers in history, the greatest accumulation of plastic water bottles in history, the greatest accumulation of golden parachutes in history, the greatest accumulation of antibiotic-resistant strains of bacteria in history, the greatest accumulation of divorces in history, the greatest accumulation of nuclear warheads in history, the greatest accumulation of chemicals in the soil and water in history. We are drowning in sound bites, instant messages, stock quotes, and advertising. We devote infinitely more time to watching what Carlo Petrini calls "food pornography" on TV than we do to cooking. We are, in the words of a recent émigré from the Middle East to England, "awash in sexuality." (Pamela Anderson's Web site is, according to Iventa, "the most sought after property on the Internet, at times generating over 180,000,000 searches per day.")

Ever since the first sheep commons was enclosed in England, during the early days of the Industrial Revolution, the extent to which economic growth depends on disruptions to established social and environmental relationships has been evident. Granted, the greatest legal accumulation of wealth in history can mask a good many ills. But only for so long, and "so long" is coming.

September 11 was one of its precursors. The War on Terror is a precursor. Peak Oil and 380-plus parts per million of carbon in the atmosphere are precursors. Suburban sprawl and trophy homes are precursors. Grand Theft Auto video game and lawsuits alleging it incites teens to murder are precursors. McDonald's and Coca Cola are precursors.

Rather than looking for enemies to blame and lashing out at the messengers who are bringing us, however horribly and inappropriately, the bad news, we need to muster the strength to look at ourselves and, so, to "see the world in new ways."

It is to my own piece of land that my attention ultimately turns.

From here, high on the llano at 8,300 feet, I can see more than 50 miles in many directions. I can see Los Alamos nestled against the Jemez Mountains, where, at night, its lights make a streak of electric glitter against the black terrain. But it is the vision of a household that brought me here.

I have come here in a continuing, highly imperfect, conundrum-riddled quest to find *here*. Another move, another hope that this time I will find it, and stay put. If it is true that Americans move on the average every five years, then I am pretty much in line, having moved seven times since 1975. Against my "footlessness" (to use E. F. Schumacher's phrase), I would hold the standard that I have posed for myself: Leave your place more beautiful than you found it. Leave your land more fertile than you found it.

We Americans are all—migrant worker and wealthy person, alike—displaced by the modern economy. Following the money or chased by the money, pursuing economic growth or pursued by economic growth or enjoying the fruits of economic growth, our allegiance to place is diminished or destroyed.

My arrival to this place has been part of a kind of cultural decompression: from an affluent restaurant-and-gallery-riddled seaside resort community in New England to a non-gentrified, abandoned-building-riddled, rural, sky-and-land-oriented mountain village in New Mexico. The money economy has ripped the heart out of both communities, but here, unlike in the resort back East, it has not resuscitated the patient and implanted an artificial heart.

Village culture here, with its local economy drained by jobs in Los Alamos and Santa Fe, seems on or near its deathbed. There is barely a town center, the general store has a For Sale sign on it, and the part-time post office recently closed, until a new location and a new postal worker could be found. Yet, there is sort of peace and stability here, as if, like a dinghy sunk in a hurricane, the town has hit bottom and is resting there. The possibility remains that one day it will be raised and restored, and that it will not, otherwise, before such time, be made over in the rush to create a world-class mooring for trophy homes. Think of them, these impressive, many-bedroomed, many-bathroomed, well-appointed second or third homes as marooned motor yachts: They sit high and dry, monumental oddities that dwarf the year-round residences in whose midst they have washed up. Blissfully and unapologetically unaware of how out of place, how stranded, how out of touch they are, they might as well have names like Enchantment and Esmerelda II and Truth or Consequences displayed conspicuously on them like self-congratulatory, private jokes for all to see. New SUVs, parked with precious few miles on them, sit like spotless, fancy Zodiacs tied to their sterns.

I live here, as I did back East, powered solely by a small array of solar panels. I work out of my home, drive as little as possible, have a raised-bed garden in which I grow a few vegetables, and look forward to converting the old chicken coop into a composting-toilet outhouse. My kitchen was, prior to the 1970s, a log-built shepherd's pen, of which scores dot the high meadows in varying states of extreme tumbledown, monuments to an age of seasonal grazing that lasted a few hundred years but did not survive World War II.

The disparity between the kind of household toward which I aspire and the culture represented by trophy second homes is striking.

It occurs to me that only two kinds of cultures celebrate multiple residences and trophy homes: the unabashedly aristocratic and the unabashedly capitalistic. Kings have summer palaces. Princes have rural estates. The upwardly mobile have second homes. Captains of industry have third homes. Captains of finance have fourth homes.

To which a late-sixties Firesign Theater album answers, singing in a kind of goofy jingle style, "*How can you be two places at once when you're not anywhere at all?*"

Wendell Berry has his own version of *not anywhere at all*:

> Those are the two poles between which a competent morality would balance and mediate: the doorstep and the planet. The most meaningful dependence of my house is not on the U.S. government, but on the world, the earth. No matter how sophisticated and complex and powerful our institutions, we are still exactly as dependent on the earth as the earthworms. To cease to know this, and to fail to act on upon the knowledge, is to begin to die the death of a broken machine. In default of man's personal cherishing and care, now that his machinery has become so awesomely powerful, the earth must become the victim of his institutions, the violent self-destructive machinery of man-in-the-abstract. And so, conversely, the most meaningful dependence of the earth is not on the U.S. government, but on my household—how I live, how I raise my children, how I care for the land entrusted to me.
>
> . . . To assert that a man owes an allegiance that is antecedent to his allegiance to his household, or higher than

> his allegiance to the earth, is to invite a state of moral chaos that will destroy both the household and the earth.[3]

To choose homeland over household is to be *not anywhere at all*, lost in political abstractions and suffering from a kind of "spiritual nomadism" that is the shadow side of the frontier mentality, the sense of Manifest Destiny that has been so identified with American greatness.

Conquest is one thing, settling another. That the American people have an aptitude for conquering is beyond doubt. We have conquered not once, but many times, not only the land between the seas, but also the space between the planets. We have vanquished political and military foes. That we, amidst all these victories, yet have failed to *settle* is something that seems much less well understood.

Consider the litany of uniquely American attributes: pioneering spirit, entrepreneurship, technological innovation, Wall Street, the land of opportunity. They are all about movement, about looking forward and taking risks, about fortunes to be made and breakthroughs to discover. There is nothing in them about permanence or stability or patience or community.

And, so, fail to settle we have, leaving behind us a landscape of tear-down buildings and temporary strip malls, abandoned urban centers and suburban sprawl, towns left for cities and cities left for suburbs and suburbs left for vacation homes and hometowns left in pursuit of the good life, which is always just over the horizon, linked to a company or a profession or a career opportunity or an investment opportunity or a booming

3. Wendell Berry, "Some Thoughts on Citizenship and Conscience in Honor of Don Pratt," in *The Long-Legged House* (New York: Harcourt, Brace & World, 1969).

market. We have chosen allegiance to markets over allegiance to place and to household, masking our choice in the politics and economics of export: We export grain, we export cigarettes, we export movies, we export technology, we even export democracy. We have become experts at creating and serving markets, but we are failing to maintain the fertility of the social relationships on which culture depends.

"To waste the soil," Berry continues in his essay "Some Thoughts on Citizenship and Conscience in Honor of Don Pratt," "is to cause hunger, as direct an aggression as an armed attack; it is an act of violence against the future of the human race." This essay, written in 1969 after the arrest of a University of Kentucky student protesting the war in Vietnam, powerfully elicits the relationship between place, politics, and morality:

> I am struggling, amid all the current political uproar, to keep clearly in mind that it is *not* merely because our policies are wrong that we are so destructive and violent. It goes deeper than that, and is more troubling. We are so little at peace with ourselves and our neighbors because we are not at peace with our place in the world, our land. Until we end our violence against the earth—a matter ignored by most pacifists, as the issue of military violence is ignored by most conservationists—how can we hope to end our violence against each other?

My efforts to settle in the mountains of northern New Mexico, and to create a household that steers away from consumerism and toward the small of scale, the nonpecuniary, and the ecologically sensible, is a living monument to one of my favorite aphorisms: *The perfect is the enemy of the good.*

The perfect is the enemy of the good. The gopher is the enemy of the garden. Distant markets and cyberspace are the enemies of the household.

Distant markets and cyberspace do not attack us directly. They come bearing gifts: food out of season, cheap goods made in foreign factories, 24/7 opportunities for shopping and surfing the Internet. But something happens as their influence pervades the household. The balance of our attention shifts from here to there. We lose touch with the notion of the household as a locus of production and settle into the notion that a household is merely a locus of consumption. Even worse, we slide unawares into the belief that the health of a household and the society of which it is a part are defined by level of consumption: Greater consumption equals greater well-being. Finally, we are seduced into paying more attention to the computer screen and the TV screen than to what is happening in our own backyard.

We choose the screen over gopher and gopher hole, the screen over swallow and swallow's nest, over the migration of the celestial orbs, over the plaintive wail of a gentle, impressive pre-dusk breeze rushing down from the peaks and past the eaves, the screen over the general store that is on the market and stocking only a skeleton stock—beer, motor oil, chips, a smattering of canned goods, and a few dozen basic drugstore items—as if to symbolize the life that is trickling out of this place, the screen over the demise of farms, the encroachment of macadam, the loss of quiet, over the departure of our children in search of educational degrees and better paying jobs. We see what we see around us every day, but we do not really believe it unless we see it on the screen.

We choose the Internet. Swirling in cyberspace, uncountable disaggregated sound bites create a democracy-mimicking smorgasbord-smog of facts and fictions, an illusory magnetic field for our attention. Seductive is this illusion that we can browse our way, effortlessly consume our way, one byte at a time, one instant message at a time, toward knowledge, and, eventually, if we snoop around tirelessly enough and fast enough, all the way to wisdom. (Secretly, we surmise that sound bites are to wisdom what empty calories are to nutrition. They create a craving. They seem to fill us up. Their verisimilitude of plenty—everything you could ever want to know about this or that, up-to-the minute coverage of this or that—gives a temporary rush.)

We choose intermediaries. We give our wealth, the creation of which is the central organizing principle of our work and living patterns, to invisible financial intermediaries, asking virtually nothing of them but that they minimize risk of loss, maximize financial gains, and ensure that we can reach back into the virtual black box for our loot whenever we might need it, at a moment's notice. Beyond the accomplishment of abstractions called *diversification* and *liquidity*, we do not care where our money goes, what it is invested in, what activities it encourages, what natural resources or communities it damages. In fact, we are taught *not* to care about these things, that these things are best left to fiduciaries and activists and other experts, that to interject such concerns into our own investments is to interfere with the efficient working of capital markets, whose efficiency is fundamental to the continuing vitality of economic growth, to which arrow our fate is tied.

We choose TV. We would rather watch *Touched by an Angel* or *Nova* than stand in the driveway staring up at the Milky Way.

We choose media. We discount the information from our own

senses and experiences, or from our neighbors, or from our local community. We turn instead to the endless and vaguely reassuring chatter of the Internet and the information provided by media channels whose top of the funnel seems enormous, but is actually quite narrow: global disaster of the day, global scandal of the day, global terrorism of the day, global celebrity highlight of the day, global economic anomaly of the day, along with today's prices of everything.

We choose convenience and speed. We drive here and fly there and consume products expressed to us from all over the world, without a thought to the degree to which our so doing is subsidized by cheap labor, subsidized by villages left behind, subsidized by pollution, subsidized by aquifers depleted, subsidized by collateral damage, subsidized by species lost, and by that which it has taken natural systems tens of thousands of years to provide.

The first step in the process of our recovery is an unabashed acknowledgment of the extent to which how we live depends on subsidy. Followed by a commitment not only to "see the world in a new way," but also—however imperfectly—to begin living in a new way.

Gophers are not the harbingers of fertility that earthworms are. They're varmints, they're critters, and they can wreak havoc in a garden, mocking our efforts toward domesticity. Even so, they and their holes represent something stubborn and real about what it takes to wrest subsistence from a piece of land.

The gopher hole is as good a place as any for us to seek grounding. It is far more immediate, far less threatening, far less abstract than the ozone hole over Antarctica or the dead zone in the Gulf of

Mexico. The gopher hole provides an antidote to that overblown and misguided talk of patriotism and self-preservation that marks the fifth anniversary of 9/11: "For the first time in my life, I know how it feels to face an existential menace. They want us to die. All of a sudden I realize what my parents were talking about all those years. Patriotism, I now believe, isn't some sentimental, old conceit. It's self-preservation."

Is the War on Terror really about self-preservation? The actual threat to our national sovereignty is negligible, and as individuals we are far more likely to die in a car crash than in a terrorist incident. This does not diminish the existential terror of the possibility, however statistically small, that a bomb could explode, any time, any place. But there is another, more insidious horror. Behind the War on Terror lurks the horrific realization that what we call the American Way of Life, iconic to the entire world, it seemed, throughout the nineteenth and twentieth centuries, is losing its luster and its moral fiber. These qualities can neither be restored nor sustained by military means.

Equating military might with moral leadership is a bit too much like confusing killing gophers with being a good gardener.

Bingo (short for Bingo Pajama, his namesake a character in Tom Robbins' *Jitterbug Perfume*) is mostly blue heeler, and although I had not expected him to distinguish himself as a terrier-like ratter, he's definitely got more than a little rodent hunter in him. He pounces in the tall prairie grasses. He digs. Occasionally, he reaches his paws under a boulder. Every now and then, there he is with a dead gopher in his mouth. I am happy to know that one less gopher is around to attack my garden. But that does not make me a booster of the Gopher Killing School of Garden Management. In defense of my garden, I do not consider killing gophers by mechanical or chemical means an option.

I suspect that there is a little Hindu in every organic gardener, since one of the basic tenets of Hinduism—do no harm—is central to the motivation to grow food organically. It is often surprising to those who know little about farming to learn the extent of the harm caused by industrial agriculture.

Industrial agriculture has certainly been a marvel of the modern world, but the harm it causes has been masked by a combination of immature environmental awareness (it wasn't until the 1960s, for instance, that the hazards of DDT became apparent) and the dramatic ability of modern farming and food-processing techniques to fill supermarkets. Over time, however, the negative impacts of industrial agriculture have become more apparent, including pollution of groundwater, aquifer depletion, soil erosion, and loss of biodiversity on the environmental side, and flagging small farms and rural towns on the social side. Modern agriculture is not about the health of the soil or the health of farm families; it is about how much food can be wrested from the soil, using a muscular formula of N-P-K chemical fertilizers, a soup of pesticides and herbicides, and production techniques that depend on large-scale monoculture, mechanization, and cheap labor.

Intensive application of a few chemicals is an expedient way to boost yields for a few decades; however, it is not a good way to replicate or nurture the complex web of symbiotic relationships that produced soil fertility over millennia. We should not be surprised that megadoses of a few synthetic compounds have deleterious long-term effects. We should not be surprised that what results from such an approach are not only mountains of commodity foodstuffs, but also degraded households, degraded communities, and degraded soil. It took millennia or eons, depending on how you count, to create fertility; we are well along the way toward materially damaging it in a matter of a half century.

The damage caused to natural systems and the fabric of rural society by industrial farming is one thing. The damage caused to my small garden by varmints is another.

There is little as challenging to the organic gardener as varmints. Deer, rabbits, woodchucks, rats, gophers—in a small garden, the damage they cause can be catastrophic. The military option is tempting, but not particularly practical. In the case of gophers on this northern New Mexico llano, there are way too many of them and their burrows are extensive. Gopher traps are not terribly effective. Poison poses questions of collateral damage, however slow and indirect. A few Jack Russell terriers would have a field day,[4] but the consensus of those who've lived around here a long time is that Jack Russells won't survive the coyotes. So, I opt for the diligent installation of rodent wire around the foot of my garden fence and under each of my double-dug, rough-hewn-lumber-sided raised beds.

The security of the garden may be briefly defended through violent means, but the health of the garden over time will ultimately depend on the health of its soil and the stewardship of its tender, which stewardship must include a sustainable relationship with all critters—call it détente, call it an ongoing dance of

4. You'd almost think that the very definition of "having a field day" would be "having as much fun as a Jack Russell terrier in a field full of gophers." No. The phrase is of military derivation, referring to parade days, which were much easier on soldiers than their customary activities on the base.

defense, call it Bingoism, call it the poetry of good fences, gumption, humility, and forbearance.

Forbearance is, perhaps, a good place for us to stop.

"What sets us apart from other species," writes Michael Pollan in *Second Nature*, "is culture, and what is culture but forbearance?"[5]

Which reminds me of something Wendell Berry once remarked, referring to Farmers Diner, a start-up company aiming to create a chain of diners that source food locally and organically, "What will make Farmers Diner remarkable is knowing when to stop."

Lacking forbearance, we of the modern economy and we of the greatest accumulation of wealth in history too often become the butt of our own jokes: "Why does a dog lick its private parts?" "Because it can."

Too often, we use force, we resort to power and technology, because we can. We spray pesticides because we can. We apply ammonia nitrate because we can. We ship food thousands of miles and fill it full of chemical additives because we can. We create companies with thousands of virtually identical outlets because we can. We grow millions of genetically identical hogs in factory farms because we can. We turn millions of bushels of corn into millions of gallons of high-fructose corn syrup because we can. We feed 65 percent of our grain to livestock because we can. We produce more and more food with fewer and fewer, larger and larger farms, because we can. We poison gophers because we can. We connect ourselves in a global electronic web because we can.

---

5. Michael Pollan, *Second Nature: A Gardener's Education* (New York: Atlantic Monthly Press, 1991).

As a result of all of this—this globalization, this creation of the so-called global village—we are enjoying unheralded benefits of being *connected*. Yet, paradoxically, we are also suffering the consequences of becoming increasingly *disconnected*.

The typical suburban home has been compared to a hospital patient on life support: water, electricity, heating fuel, and food come from far away and waste is taken somewhere far away. The household has become a place of consumption only; we produce little or nothing where we live. The dwelling place is disconnected from natural systems, dependent on distant production and complex distribution infrastructure and purchasing power.

If it is true that a house is not a home, then it is also true that a homeland is not a household and a market is not a community and a corporation is not a farm and an economy is not a place to live. Defending a homeland is no substitute for creating a nation of healthy households. We are a nation of houses and markets, but we have fewer and fewer healthy households. We are a people dependent on life-support systems from distant places, and, increasingly, on simplistic military-industrial solutions to complex social and environmental problems. We are wired and piped and highwayed and military-based all over the world—yet, we are oddly as disconnected as ever, perhaps more disconnected than ever, from the original, ennobling vision on which the American experiment depends.

Look out on this landscape. Step out into this landscape. Listen.

The mountains say, "*There is no hope in military victory.*" The sky says, "*They who forget the cycles of the sky will know only the cycles of violence.*" The land says, "*The path to peace does not lead through the battlefield.*" The desert says, "*The love that is to be born out of hate will never be born.*"[6] The river says, "*What you see as a palace is running water.*"[7] The coyote howls, "*The only war that matters is the war against the imagination.*"[8] Now, before this sounds too much like an imitation of Chief Seattle gone soft (although the mention of Native Americans does bring to mind my favorite T-shirt, showing Geronimo and several warriors brandishing rifles and bandoliers, over the caption, "Fighting Terrorism Since 1492"), let us also note that the gopher says, "*Defend your territory? Hah!*"

I pause from my writing to push back the curtains that shade my desk for the first hour or so every morning, when the sun is low enough in the east to make it uncomfortably bright in my office. Truchas Peak, the Wizard of Weather, is revealed. This morning, "He" is twirling a few medium-sized (in mountain terms) cumulus clouds.

Living with a small array of photovoltaic panels on my roof, I am conscious of the fact that September 21 marked the end of "long sun" season. From then until June 21 is "short sun" season.

---

6. Actually, this is Thomas Merton (*The Collected Poems*), who also wrote: "The real trouble with 'the world,' in the bad sense which the Gospel condemns, is that it is a complete systematic sham, and he who follows it ends not by living but by pretending he is alive, and justifying his pretense by an appeal to the general conspiracy of all the others to do the same. It is this pretense that must be vomited out in the desert." *Confessions of a Guilty Bystander* (New York: Doubleday, 1966).
7. Actually, this is Dogen Kigen, from the essay "Sansuikyo," in "Mountains and Waters Sutra," written in 1240, as cited in Gary Snyder, *The Practice of the Wild* (New York: North Point Press, 1990).
8. Actually, this is Diane di Prima from her poem *Rant*, as cited in Gary Snyder, *A Place in Space* (Washington, DC: Counterpoint, 1995).

What could be more beautiful than being conscious every day, in however small a way, of the earth's relationship to the sun? What could be more sad, and, ultimately, more destructive, than being disconnected from this most fundamental of relationships?

The new fence around my garden is almost complete. All of the rough-barked fir posts and aspen rails are in place, as is the handmade aspen gate. All of the rodent wire under the raised beds and along the foot of the fence is in place. Most of the chicken wire that fills the gap between the lower and the upper rail is also in place—after all, sometimes we even need fences to keep our friends out. (This means Bingo, who is fond of digging in the raised beds and chewing plastic irrigation lines.)

I do my best to weave forbearance into the fabric of the household that I am striving to create. There may be no true wilderness left—every nook and cranny of every continent may have been visited and picked over and mapped and mined and affected by humans near and far—but there is still plenty of need for pioneering spirit right here, in our own backyards, in this place of too many *not-anywheres-at-all* called America.

If we can exercise forbearance, we may turn our energies to the true task at hand: the restoration and preservation of the commons, while nurturing new sources of cultural and commercial vitality for the entire human family. Challenges posed by global warming, declining biodiversity, terrorism, militarism, and other systemic threats to the commons can be seen as both impending disasters or as once-in-the-history-of-the-planet opportunities for humanity to show our truer colors, to find a new way to see the world.

Such is the task to which *terroiristes* around the world can lend their complete allegiance.

SIX

the pursuit of zero

Ground Zero. Since 9/11, the term has taken on new currency. It is hard to imagine that it might betoken something other than the site of terrorist events. Yet it is precisely this notion that I explore today.

For it seems to me on this last day of October 2001, this day of the dead, as I gaze out from my electronic cottage upon the last vestiges of this year's vegetable garden, that ours is the Age of the Zero. Racing from millions of instructions per second to billions of instructions per second, from millionaires to billionaires, from horsepower to nuclear power, from Somewheres measured in seasons and lifecycles to Nowheres measured in nanoseconds and cyberspaces, ours is the age that mistook the pursuit of happiness for the pursuit of zero.

Ground Zero.

Zero: the language of instantaneous communication and unprecedented wealth and exponential growth. Zero: the disconnect that lies at the heart of the modern or postmodern or post-industrial or pre-post-whatever-comes-next culture—a culture that divorces home from work, production from consumption, common sense from technical know-how, information from understanding. Zero:

the face of affluence, invisible behind the veil of free markets. Zero: the zip code of Anywhere USA, a culture that is so homogenized and ultrapasteurized that raw milk is virtually unknown.

The twin towers of Money and Speed are built not of steel, but of Zero.[1]

Zero is the stuff of Other People's Money, the stuff of financial institutions so huge and financial markets so fast and so vast that even the tiniest decimal point yields a Christmas bonus that no fiduciary can resist.

Zero, it is, that turns us all into investors.

And after all the Zeroes . . . what then? We come back to Ground.

Ground: both the ground of our being and each actual piece of ground, which, taken with all others, comprise the planet from which our sustenance derives. Ground: to which all aircraft, and all

---

1. In his 1961 poem entitled "The Moslems' Angel of Death," Thomas Merton wrote:

> The firefly city stirs all over with knowledge.
> His high buildings see too many
> Persons: he has found out
> Their times and when their windows
> Will go out.
>
> He turns the city lights in his fingers like money.
>
> No other angel knows this one's place,
> No other sees his phoenix wings, or understands
> That he is lord of Death.
>
> (Death was once allowed
> To yell at the sky:
> "I am death!
> I take friend from friend!
> I am death!
> I leave your room empty!")
>
> O night, O High Towers! No man can ever
> Escape you, O night!

markets, and all home-run balls, and all living things, must ultimately return. Ground: the place where money lights, where money rushes in and rushes out, leaving megalopoli and malls, dumps and gated communities and cul-de-sacs and barrios and foreign debt, a few hundred billionaires and a few billion hundredaires, a commercial cornucopia of consumables, supermarkets filled with commodity foodstuffs: food designed for long shelf life and filled by food technologists with preservatives and flavor enhancers and empty calories, food grown by farmers who must be agribusinessmen first, stewards, second, if at all. (Zero would appear to be the number of farmers toward which the industrialization of food production is taking us.)

And then it hits me, as if the poet-zealot in me hijacked my more prosaic thoughts and flew them straight into the place reserved for finance: The real ground zero, the ground zero on which the future depends, is not the former site of the World Trade Center. Rather, it is the place where money meets nature: Ground zero is the farmer's field, the vegetable garden, the place where every day the battle between economy and ecology plays out through the process of agriculture.

"Our distance from the source of our food allows us to be superficially more comfortable, and distinctly more ignorant," writes Gary Snyder. "Our stance in regard to food is a daily manifestation of our economics and ecology. Food is the field in which we daily explore our 'harming' of the world."[2]

If we cannot grow food in a way that leaves the soil as fertile or more fertile than we found it, if we cannot grow food without leav-

---

2. From "Nets of Beads, Webs of Cells" and "Survival and Sacrament," in *The Practice of the Wild*. The word *harming* is in quotes because it is part of an exposition on the Buddhist principle of *ahimsa*, the principle of committing no harm.

ing biocides in our water, if we cannot cultivate without destroying both biodiversity and the fabric of rural communities, if we cannot grow food without being driven by the imperatives of the largest scale, lowest cost, export-oriented, extractive, industrial production system, what does this say about the health of our society, about our true wealth, about the prospect for future generations?

We are not only what we eat, but how we grow what we eat.

Food is Ground Zero.

I did not have these thoughts in a vacuum. They ruminated over the weekend, while I was visiting Eliot Coleman and Barbara Damrosch at their farm in Harborside, Maine, where they grow organic vegetables throughout the winter in a way that is as much a work of art, a passionate dance, a radical departure and return, an obliteration of the line between epicurean and peasant, as it is a greenhouse production system. Eat their carrots. (Admire them first.)

While I was sitting at their breakfast table, it occurred to me: The task at hand is to create a portfolio of venture investments in early-stage sustainability-promoting food companies that is to a traditional venture capital portfolio what Four Season Farm is to a Dole pineapple plantation.

I wish, therefore, to propose a new kind of fund.

I don't know if it would be called the Ground Zero Fund or Food One or 1,000:1 Capital[3] or Hornworm Moneymakers or something else. And I don't know the details of how it would be orga-

3. After Eliot Coleman, as previously cited: "Want return on investment? At 1,000:1 in four months, a tomato seed makes even the highest fliers seem paltry."

nized, except to say that it would not be organized like a traditional ten-year venture capital partnership, which is designed around the goal of producing 20 percent-plus internal rates of return.

I do know that it would be driven by principles, by mission, by the imperatives of nature rather than by the imperatives of finance. Its first principle would be, I suppose, the principle of carrying capacity, embedded in a process of *nurturing*, rather than *exploiting*. We must return, again, to that seminal passage from Wendell Berry's *The Unsettling of America*, which outlines the difference between exploitation and nurture: "The exploiter typically serves an institution or organization; the nurturer serves land, household, community, place. The exploiter thinks in terms of numbers, quantities, 'hard facts'; the nurturer in terms of character, condition, quality, kind."

I read this passage at an Investors' Circle meeting a few years ago and the audience's resonance with it was almost palpable. Yet try to translate these concerns into strategies for investing money in a more nurturing way, and that resonance turns to dissonance in a . . . nanosecond. For the vast majority of even this rarified, self-selected universe of sustainability-seeking high-net-worth investors, maximizing financial return is axiomatic. Do well while doing good. Maximize IRR *and* ERR.[4] But if *nurturing* means, either directly or indirectly, "sacrificing financial returns," then all bets are off.

Since internal rate of return is extremely sensitive to time (as in, the shorter the holding period per given dollar earned, the higher the rate of return) and since natural systems work in seasons and eons rather than in quarters and years, it seems axiomatic to me that internal rate of return is a fundamentally inappropriate

4. External rate of return, as described in Chapter 1.

benchmark by which to measure the success of a portfolio of early-stage organic, sustainable, or slow food investments. Or, stated in the reverse, professional venture capital, and its return objectives, have evolved around a very narrowly defined universe of high-tech start-ups that have extraordinary, explosive growth potential.

Let us design an investment vehicle that seeks not to maximize speed and power, but rather to optimize the health of human communities and natural systems. Let us affirm that we cannot continue to invest as usual, content to use the income from those investments to address philanthropically the problems created by the commercial enterprises in which we have invested. We need to start trying to fix the problem at the front end of the pipe.

We must forget about the distinctions, as defined by the Internal Revenue Service, between for-profit and nonprofit. We must even forget about the possibility that there may be such a thing as ERR or SROI or blended returns—efforts to quantify the "nonfinancial" or "exogenous" impacts of our investments, the social and environmental "returns." We must act from a different place, a place that is at once deeply compassionate and militantly nondogmatic and stubbornly pragmatic.

Our departure on this new course would seem to be fraught with imponderables: Can money be used in such a fundamentally different way? Can investors reimagine their relationships to companies? Can companies reimagine their relationships to nature and to markets?

We are not intelligent enough to see beyond this horizon. We need not.

Let us be heartened by imagining that a seed has no consciousness of the fruit into which it is going to evolve. (And in one short season.)

I confess that I am not really a venture capitalist.

I confess that my primary allegiance is to poetry and to nature, not to money. I confess that my heroes are Wendell Berry and Scott Savage, not Bill Gates and Jeff Bezos.[5] My hormones seem stimulated more by the vibration of epiphany than by the scent of profit. I confess that I believe that in today's world it takes more strength to go slow than to go fast, more integrity to swim upstream against money than to stay out in the relatively anonymous depths of the ocean of global markets. I confess to lying awake at night wondering about the hypocritical aspects of my dependence on distant financial institutions and the inexorable violence that arises out of unbridled allegiance to political, religious, and economic ideology. I confess to utter frustration at the taboo of speaking openly about money: Secrecy about these matters is the final frontier that stands between us, as we are today, and us as we need to become to leapfrog the destructiveness that we have unleashed. For all our technologically enhanced peering—peering out into space and peering down into the atom and the gene—we are blinded, in our daily life choices, by money.

---

5. Scott Savage is an Amish farmer and former director of the Center for Plain Living in Barnesville, Ohio, where he published *Plain Magazine* and *The Plain Reader*: "If information highways are the wave of the future then I will build information country roads on which the traveller can reach the truth faster by going slower. . . ." One reviewer commented, "Reading about a garden cooperative in Connecticut, the raising of a home with only plaster and straw in hand, a fascinating trip to New York City through Amish eyes, compels each of us wonder: Can I too survive without television or that high-tech appliance cluttering my kitchen counter? Am I just a cog in the wheel of the global economy? Is isolation from one another and from the earth the simple destiny of humankind? Each rich, personal essay in this provocative collection offers solace, wisdom, joy, and quiet space for contemplation."

I confess, therefore, to being a *nurture* capitalist.

I also confess to having mused on more than one occasion: Might the world ever see an annual report written in verse?

Doing my bit to hasten that day, I splice herein a poem.

Twin Lights

I

At the head of Gloucester harbor
they stand, on a small island of rock.

II

Posts to no beam, they seem,
of a structure never to be built:
Twin Lights.

Dual exclamation points, fore and aft
of no sentence, a pleasing redundancy,
if a bit puzzling::

goal posts for some Poseidon.

(Did he swim these waters?
Or was it some Indian totem
with a name like Annasquatum
or Pokanucket: Oh, if they had only
built temples instead of wigwams!:)

III

Wait: Did you catch that?

That: **! :**

We must ask the question, then:
Can a colon stand beside an exclamation point
like that? To what end? Like two odd markers
in their own right, too close together to
serve any real purpose, aren't they,
to provide any direction across the sea
of thoughts upon which words would sail?

: ! That!

IV

We must ask the question, then:
What are they doing there together, Twin Lights?
Monuments to a mercantile mind,
a schooner fleet, once so grand,
as grand as the Grand Banks, indeed,
that no expense was spared
to ensure their safe return?

V

They make a solitude of two.
They make a portal for the mind to pass through.

VI

Imagination awakens just before dawn.

As if sensing the intensity of the light
about to pour in through the picture windows
after the curtains are pulled wide, she waits
with a kind of jocular casualness, a sleepy-eyed
inquisitiveness, for History, her breakfast mate,
to rouse in the next room, to peer at the sun.

VII

History and Imagination,
rock and water,

apposite, defining, unyielding to our eye,
making a place for the sun to rise over,

for moon-driven tides to carry uncountable cargoes
in invisible hulls, harbor bound, past Twin Lights.

VIII

At first blush, fishing and imperialism
have little in common.

One is a net-flinging, a diurnal coming and going,
a harvesting. The other a subjugation,
a claim-staking over generations, a banking.

But fishing has itself been netted, hauled aboard
by a commerce of diminishing returns. This dying
harbor town, an end point for an exhausted logic.

When fishing became an industry, it passed
through a portal through which there is no return.

IX

At first blush, poetry and capitalism
have little in common.

Poetry is a recapturing,
a return past History
to a vast darkness
through which men peered
with far less certainty,

forever coming up against
the limits of their own eyes,
forever entering into
a negotiation of predawn seas,
a calculation of sextant-souls,
a business of stars and senses
and winds and tides, of great ships
and infinitely greater seas,
a commerce of all you could haul,
a world with a far fishier future.

Poetry is the portal
through which capitalism can return.

Now, just as History will not yield in any final way to Imagination, we can rest safe in the knowledge that Economics will not yield in any final way to a new, more poetic, ethically correct vision of

Ecology. If it did, I suppose, something called *ecopoethics* might be the result.

In the meantime, those of us who are trying to self-organize around a new way of investing need new, more poetic ways of thinking and speaking about how we are trying to use our capital.

Eliot Coleman writes:

> In the opening paragraph of his classic *Soil and Civilization*, Edward Hyams decries how modern misapplication of science has caused humans to "begin working across or against the grain of life." Hyams notes how science, when it becomes master rather than servant, displaces age-old natural wisdom that has maintained the "integrity of life." Without that integrity, humans begin to lose contact with the "poet," which Hyams describes as the instinctive understanding of wholeness that has nurtured their well-being through the centuries.
>
> Such change is abundantly evident in our modern American diet. The business of food science is in conflict with the poetry of human nourishment. Store shelves are filled with products that keep seemingly forever, such as canned or frozen food, ultra-pasteurized dairy products, devitalized flour. . . . My instincts tell me that long-dead foods cannot properly nourish long-lived people.

I would like to invite the poet to sit down with the investor, that we might arrive at a new way of thinking and speaking.[6] A way that is distinctly less prosaic and distinctly more playful, more

6. Wallace Stevens wrote, "Money is a kind of poetry." What in the world did he mean by that?

humane, more welcoming of the mystery of the nonhuman, less utilitarian, more immediate than the pseudoscientific jargon of finance. For the new course of action we seek depends on a new articulation.[7]

We will not find our way to more humane, ecologically sound, socially just corporate cultures so long as we continue to use outdated language. Bottom Line and Invisible Hand and Prudent Man were useful metaphors for a century or so, while the global economy was springing from the world of a billion global inhabitants (human, that is) and the Model T. Using the same metaphors in today's six-billion-person, cyber-surging world makes us a bit too much like Roadrunner in the old cartoon series: We are over the edge of the cliff, spinning faster than ever, hoping that, somehow, if we do it hard enough, we'll be able to avoid falling. But there is no cartoonist to save us from the abyss of terrorists and toxics.

Given the urgency of the hour, there is no time to rest contented with the language of portfolio managers and financial advisors. We need to choose *direction* over *diversification*. We need to approach our role as sustainability-minded investors with a determination to find ways to say yes, knowing that there are always nine reasons to say no.[8] We need to speak more directly from the heart, getting beyond the defensive posturing into which social investment

---

7. Wallace Stevens wrote, "The significance of the poetic act, then, is that it is evidence. It is instance and illustration. It is an illumination of a surface, the movement of a self in the rock. Above all it is a new engagement with life." From *Opus Posthumous* (New York: Alfred Knopf, 1975). If, as he also observed, money is a kind of poetry, then it will also need to be informed by a movement of the self in the soil. This is our direction: towards money as a tool for a new engagement with life.

8. The old saw asks: What's the difference between a manager and an entrepreneur? A manager has an idea, takes it to his immediate superior, and if it is approved, it goes up

discourse is always pushed. We are defining a new *kind* of investing, and we need to find ways of talking about it that are proactive, creative, and courageous.

We need to have the courage to venture, with our investing selves, out from the left side of our brains.

Perhaps ground zero is the place where the right side of the brain collides with the left side.

This was the ground zero that my son, Zander, then six years old, struggled with during Operation Desert Storm.

"My brain tells me that war is wrong. But some other part tells me that this war is right. I guess the answer is somewhere in the middle."

A few days later, as we lay in bed together, he said, "Sometimes there is a war in my brain. One side says 'Do it.' The other side says, 'Don't do it.' The middle part of your brain is the smart part. The middle part of your brain is where the answer is."

The dissonance between the ideal of nonviolence and the reality of war is only the most extreme, most visible manifestation of a dissonance that is with us everyday, the dissonance between the changes we are trying to effect "out there" (through political action, advocacy, philanthropy, or investing) and how we live our daily lives.

We invest in organic-food companies but have kitchens stocked

---

the chain of command: If nine managers on the way up say yes but the last one says no, then the idea is quashed. To an entrepreneur, nine no's are the path to the one yes he needs to go forward.

with processed foods. We give money to Greenpeace but drive an SUV. We understand the risks of mad cow disease, and want something done about it, but don't think we are related to it in any meaningful way through the bacon cheeseburger on our plate.

We reckon such dissonance to be an unavoidable consequence of the scale and complexity of modern life.

Yet this reckoning is what makes violence possible. Disconnecting our personal actions from the world theater, we unleash the power of impersonal institutions to act in ways that we would never act ourselves. We let institutions do our dirty work for us, under the guises of Fiduciary Responsibility and Efficient Capital Markets, Consumer Confidence and Military Intelligence.

It would seem that in a world of stealth bombers and counterterrorism, the dissonance of our daily lives is puny and irrelevant. Yet it is only by addressing this dissonance, by beginning to reduce it, that we can begin to effect a systemic change that is the only hope of lasting peace. Peace depends on nonviolent economies. Nonviolent economies depend on nonviolent households.

Reducing violence, not maximizing growth, would be the primary goal of an economy that was truly oriented toward enhancing human well-being. Reducing dissonance, not maximizing financial return, would be the primary goal of an investment that was truly oriented toward reducing violence.

My pursuit of this particular zero—the zero of "zero dissonance"—was shaped in significant measure by my experience as treasurer of the Jessie Smith Noyes Foundation, a New York-based environmental grant maker with assets in the $75 million range. Working closely with the foundation's president, Steve Viederman, we set out to "reduce the dissonance" between the foundation's philanthropic purpose and its asset management. "Reflecting this pervasive dissonance," we wrote, "foundations often find themselves in the position of supporting with their investment dollars activities that are antithetical to the charitable purpose of their grant making. By accepting as axiomatic the iron curtain between making money and giving it away, foundations reinforce the kind of corporate culture that identifies corporate responsibility with financial management, relegating social and environmental problems to the provinces of politics and philanthropy."

The diagrams on page 187 illustrated our view of the shift from a bifurcated corporate culture toward "zero dissonance."

The challenges were legion, on many levels, from asset allocation to legal issues to board relations. The foundation had never included venture capital in its asset allocation, but deemed mission-related venture capital to be an important dissonance-reduction tool. Would investment decisions that took the foundation's program concerns into account be a breach of fiduciary responsibility? Would too much board time be devoted to discussions of finance?

The challenges of this process of dissonance reduction were highlighted by a presentation to the board by Gary Hirshberg, the CEO of Stonyfield Farm, the private yogurt company in which the foundation had invested as part of its mission-related venture capital portfolio.

I was seated next to Gary in the board room, and as he spoke, I thought, "It doesn't get any better than this. A great entrepreneur. An amazing fit with the foundation's mission." As it happened, Gary was not only an extremely capable for-profit entrepreneur, whose company was at that time passing $30 million in sales and

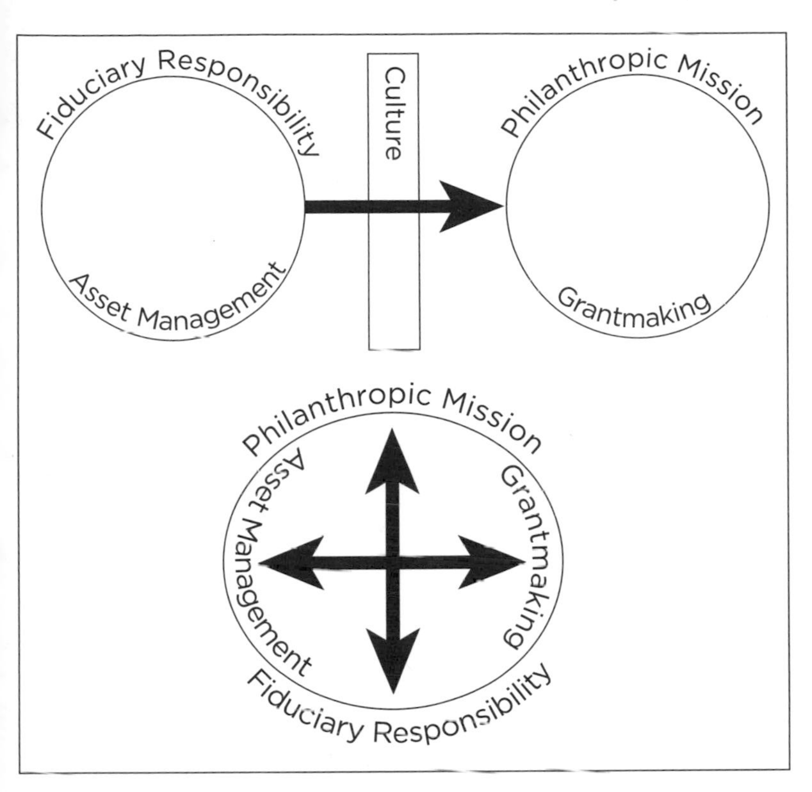

setting the pace on a host of corporate social- and environmental-responsibility issues, but he had been executive director of an environmental nonprofit that had, many years before, received grant support from the foundation. He spoke with extraordinary power about his company as a vehicle for educating consumers and for supporting small organic farmers. Yet, the morning after Gary's presentation brought not the broad consensus and enthusiasm that I had expected, but harsh criticism from several board members, who expressed their concerns that Gary's values had been co-opted, that he had become devoted to growth for growth's sake, that Stonyfield made premium products for upscale markets while the foundation's concerns for environmental justice focused on disadvantaged populations. One board member, who had come to the board from Noyes' grantee community, said, "I joined this board to fight corporations, not to spend precious board time talking about investments and corporate finance."

The dissonance inherent in the modern economic worldview is nowhere more evident than with respect to food. In the name of cheaper, more plentiful food, we are doing violence to our rural communities, diminishing biodiversity, destroying topsoil, eutrophying groundwater, and putting toxics into the food chain.

At a safe remove from the ground of this action, suburban and urban Americans find it difficult to believe that something so unsustainable could be unfolding in the heartland. At the same time, proponents of industrial agriculture point to the improvements in agricultural productivity of the last several decades as evidence against modern-day Cassandras, who have been prematurely forecasting limits to growth for several decades now.

Arguments about the sustainability or non-sustainability of industrial agriculture are not just about environmental capital, however. There are steep social costs associated with dredging the Missouri River so Midwest farmers can compete with farmers from Brazil to barge ever cheaper soybeans to the doorsteps of global processors. There are steep social costs associated with hog-confinement units, which turn pork bellies into a commodity that no family farm can afford to produce. Not just our agriculture, but our culture is diminished when farms become factories. We fill our animals with antibiotics and growth hormones; we empty our towns. We allow agriculture to leave our shores, in favor of cheap land and cheap labor in developing countries, failing to realize that we will be neither richer nor more secure as a country in which tourist destinations have supplanted productive landscapes.

Coming closer to my own ground zero, the bankruptcy of modern economics manifests itself everyday in my garden.

Consider the ten-dollar head of broccoli. If I take into account the hourly value of my time as measured by the marketplace, each head of broccoli that I grow costs me at least ten times what I could purchase an "equivalent" head for at the supermarket, or a somewhat lower multiple of the price of an "equivalent" head at the health food store. In terms of economic rationality, my time working in the garden is wasted: I am investing thousands of dollars' worth of time to produce vegetables with a market value of hundreds of dollars.

To a "ground zero" way of thinking, there is no such thing as an "equivalent" head of broccoli available from any purveyor, and what is incalculably valuable is the satisfaction that comes with good work that is connected to the land. If it is not rooted in respite from good work, leisure becomes as cheap as the cheap food that makes it possible. If it is fresh, organic, and the product of my own nurturing over a few-month period, then broccoli is something more than just a product to be valued in terms of its market price and the market value of my labor.

You must ignore fundamental, qualitative distinctions in order to accept the notion that ten heads of supermarket broccoli or five heads of health-store broccoli are equivalent to one head of homegrown broccoli. We should nail Oscar Wilde's aphorism to our garden gates: "A cynic is a man who knows the price of everything and the value of nothing."

The cost of cheap food is devalued culture.

We have before us, then, an opportunity not only to put money to work in support of sustainable agriculture, but to do so in a way that embraces broader cultural purposes. We understand that a portfolio committed to early-stage food companies will not be competitive with traditional venture capital in terms of financial return. We recognize that even within the universe of double-bottom-line food companies that is our purview, it will be important for social and environmental reasons to deploy some of our assets in enterprises whose scale or structure will further mitigate against maximum financial returns. In keeping with the principles of carrying capacity and nurture, we will explore structural alternatives to the conventional venture capital LLC, including capping investor returns and reinvesting a portion of the fund's proceeds toward our longer term vision.

Zero is our allegiance to a particular fund design or funding strategy; we must let form follow function, being willing to traverse the Demilitarized Zone between investing and grant-making, finance and activism.

Ground Zero—the one in lower Manhattan—is a symptom of a global disease. The need for actions that will begin to address this disease at its—forgive me—*roots* is profound and urgent.

Investing in sustainable food enterprises may seem an odd and ineffectual response to the world after 9/11 and dot.bomb[9] to be sure. The idea of wealthy Americans investing in the production of organic food may seem oddly out of step with the hunger that haunts the two billion people on earth who live on less than $1 per day.

Quite to the contrary, I can see few tasks that go more to the heart of our global challenges. It was agriculture that gave birth to the modern economy, and it must be agriculture that we fix, if there is to be a postmodern economy:

> The millions of years our forebears spent roaming over the landscape as hunters and gatherers left important marks on our human natures. During that period, interacting genetic and cultural evolution affected everything from our sexual behavior and religions to our food preferences. But without the ensuing agricultural revolution and the sedentary life and divisions of labor it eventually made possible, cultural evolution could never have produced our complex modern civilization and the human natures that go along with it. Without farming, which freed some people of the chore of wresting nourishment from the environment, there would be no cities, no states, no science, and no mayors, fashion models, professional soldiers, or airline pilots. . . . The agricultural revolution led to a period of cultural evolution unprecedented in its rapidity and scale. . . . It is a story that starts with the obtaining of food but returns us to

9. The term "dot.bomb" is borrowed from a book of the same name, by J. David Kuo, documenting the collapse of the Internet bubble in 2000.

> two aspects of human behavior that, although present in hunter-gatherers, became even more important in sedentary groups—religion and violence.[10]

The mention of airline pilots, religion, and violence in a paragraph about agriculture takes on a different hue in this first decade of the 21st century.

Coming full circle from a revolution that started ten thousand years ago, it is only fitting that we, as stewards of wealth, armed with the knowledge of the difference between exploitation and nurture, and believing in the power of entrepreneurship, return to agriculture as the proving ground for a new, nonviolent economy.

Ground zero, we might conclude, is not only the place where Economy meets Farmer's Field, but also the place where Investment meets Philanthropy and where industrial capitalism meets whatever comes next.[11]

This is our place. Let us tend it.

---

10. Paul Ehrlich, *Human Natures: Genes, Cultures and the Human Prospect* (Washington, DC: Island Press, 2000).

11. We are not the only ones struggling with our own version of ground zero. We are a tiny part of an enormous cultural evolution that is affecting all schools of thought, as the caterpillar of industrial economics metamorphoses into the butterfly of ecology and bionomics. Consider the following description of the tension between relativity and quantum physics: "Black holes are very, very massive, so they are subject to the laws of relativity; at the same time, black holes are very, very tiny, so they are in the domain of quantum mechanics. And far from agreeing, the two sets of laws clash at the center of a black hole. . . . Zero dwells at the juxtaposition of quantum mechanics and relativity; zero lives where the two theories meet, and zero causes the two theories to clash. . . . It is at the zero hour of the big bang and the ground zero of the black hole that mathematical equations of our world stop making sense." Charles Seife, *Zero: The Biography of a Dangerous Idea* (New York: Penguin, 2000).

epilogue

"As it circulates the globe with ever accelerating speed, money is sucking the oxygen out of the air, the fertility out of the soil and the culture out of local communities."

What started for me as metaphor has become, over a number of years, and courtesy of the goodwill and shared concerns of too many folks to name, a prescription for action.

As the preceding essays evidence, this progression has been gradual, and my steps, at times, halting.

When I first wrote "The Pursuit of Zero" and shared it with several dozen friends, colleagues, and potential investors, I knew that the odds were very slight that it would elicit real action in the form of a new fund. Emotions ran high from the events of 9/11, confusion mixed with urgency, and the requisite breadth of vision was not manifest. Yet it seemed to me then, as it does now, that certain words needed to be uttered, as if almost in a form of invocation, in order to determine if a new path might be possible.

Talk of invocation, and far-ranging references of the poetic kind, seem sometimes highfalutin' or ineffectual with respect to the formation of a new investment theory or a new fund. What could such musings have to do with the real business of managing money? More, it seems to me, than it would be prudent to attempt to calculate.

The past few decades have been studded with social investment pioneers and social entrepreneurs who launched their paper airplanes into the ether of mission-related-this and social-purpose-that and triple-accountability-and-transparency-the-other-thing, only to be severely buffeted by the crosswinds of competitive returns and the commonly accepted performance benchmarks and indices of industrial finance. If that weren't enough, once these airplanes flopped to the ground, they were trampled by the elephant in the room: economic growth.

The pressures of the market to use backward-looking metrics instead of forward-looking leaps of faith are ruthless, relentless. Without a particular strength, a particular kind of vision—sister, perhaps, to what some have called militant nonviolence and others have called pragmatic idealism (is it fiduciary activism?)—efforts to transform capital markets always seem to devolve toward the lowest common denominator, toward that fate worse than composting: doing the numbers, the macro and micro-economic numbers. Relationships continue their tumble down into transactions.

This is why we need Wallace Stevens and Jack Lazor, Wendell Berry and Joan Gussow, Paul Muller and Mohandas Gandhi, E. F. Schumacher and Eliot Coleman, and, even, Tom Robbins and Rod Serling, and the jokes of our fathers and the insights of our son. Let's not stop there. Let's get Joni Mitchell into the act before we are done: "Money makes the trees come down. It makes mountains into molehills. Big money kicks the wide wide world around."

We need to gather a new chorus of cultural voices, capable of beckoning Economic Man back to the campfire, that he may hear a new story and learn a new song.

If we cannot invoke an ignorance that is greater than our wisdom, a humility that is greater than our technological prowess and our financial proficiency, how will we find the creativity and the courage to truly use our wealth to reengage with life?

When I offered, in the Prologue, that we should listen less to economists and more to poets and farmers and the like, this was neither personal affectation nor idle daydream. John Kenneth Galbraith wrote about public policy as a necessary countervailing force to the power of the industrial corporation. Today, we are in need of something no government program and no economic formula can provide. We are in need of, to hijack one more time the words of President Bush, "a new way of seeing the world." We need a countervailing force to consumerism and market mania. Such a force can emerge only from individuals, and only when their "citizen selves," who lie awake at night worrying about air and water and justice, are reconnected with their "investor selves," who lie awake worrying about keeping ahead of inflation and health care costs and, in a few more cases every day, how much of each dollar is going up the chimney of a cement factory in some Chinese city of a million inhabitants whose name we will probably never hear on the news. It is my hope that Slow Money will contribute to the emergence of this countervailing force, joining aspirations for change, that are welling up across the land, with new ways, tangible ways, comprehensible ways to deploy capital.

In this hope I am joined by many.

We have formed a new NGO, Slow Money, and capitalized it with seed capital from 55 individuals and two foundations. The mission of Slow Money is:

- To support entrepreneurship that promotes soil fertility, appropriate-scale organic farming, and small food enterprises;
- To catalyze foundation grant making and mission-related investing in support of sustainable agriculture and local economies; and,
- To incubate next-generation socially responsible investment strategies around principles of carrying capacity, care of the commons, sense of place, cultural and biological diversity, and nonviolence.

To pursue this mission, we are convening Slow Money Institutes, regional workshops exploring strategies for investing in local food systems. We are conducting field research and creating a model slow money portfolio, from which we hope to design and launch a first Slow Money fund in 2009. We are building the Slow Money Alliance, whose members include leading food entrepreneurs, farmers, investors, and philanthropists. We are collaborating with and raising money for Slow Food, whose 80,000 members worldwide support local food communities and biodiversity.

Our work springs from many thousands of discussions and related entrepreneurial efforts. Discussions with financiers who are, wittingly or unwittingly, in search of a different path. Financiers such as my dear friend John Fullerton, who, after a career on Wall Street, came to a wonderful affirmation: "The purpose of capital is to sustain life." Discussions with farmers and entrepreneurs and foundation officers and individuals of inherited wealth and those of the Slow Food persuasion. You have but to feel the energy of Terra Madre (I am speaking, here, of the biannual event in Turin, but a more metaphorical innuendo is playfully entertained) to sense the possibilities of renewal.

This cultural renewal in the heartland of entrepreneurship and industrial finance, this biological renewal in the soil under our feet, is our ineluctable—some would call it "immune"—response to events that are unfolding near and far, as near and far as China. After millennia of civilization built around Confucian tenets of order and harmony, China lurched violently in the twentieth century first to communism and then to a genetically modified form of communism that has genes of capitalism spliced into it, creating what we face today—a high-yielding, high-growth monster that no one seems to understand. Hundreds of millions of Chinese are rushing to place their bets alongside ours.

Consider the table upon which we all wager:

circulation	vs.	percolation
monoculture	vs.	diversity
transaction	vs.	relationship
profitability	vs.	fertility

We all continue to place our bets very heavily on the values on the left side. Very heavily. So heavily that it would not seem necessary to argue the virtues of placing small counter wagers. But we have become a culture of speculators, and the longer the odds get, it seems, the more compulsively we throw ourselves at the roulette table, the more compulsively we act out our engrained patterns of gambling.

After all, what is it we are in danger of losing? Our money, which seems infinite? Our soil? What on earth could be more inexhaustible than that?

I wonder if the cycle of a single human life as described in the Bible had been called "soil to soil" whether we might have developed a more direct affinity for the life force that derives therefrom, whether, in the name of progress, in the name of modern agriculture, we would have been so ready to take what we could from it and leave the rest for just that—dust.

The September 2008 issue of *Ode Magazine* quotes John Doerr: "Energy is the mother of all markets."[1] He describes his interest in renewable energy and clean technologies in terms of trillions of dollars. In an article in the *New York Times* that same month, economic journalist Thomas Friedman defends innovation as the key to our national prosperity: "Our focus needs to be on strengthening our capacity for innovation—our most important competitive advantage."[2] He is referring to innovation of the MIT kind, innovation of the Silicon Valley kind, innovation of the "mother of all markets" kind.

This is good. This is very good. It is essential. And it is not enough. It requires, along with it, pursued with the same vigor, but with a very different sensibility, a program of healing, of nurture, of relationship building at the level of household, farm, community, and watershed. Without such relationships, all the transactions and all the innovations in the world will not yield durable wealth or well being.

---

1. Justin Mullins, "The Six Trillion Dollar Men," *Ode Magazine*, September 2008.
2. Thomas Friedman, "Georgia On My Mind," *New York Times*, September 7, 2008.

To say this makes me neither a Luddite nor a Marxist. (Although I was intrigued to find in David Montgomery's recent book, *Dirt: The Erosion of Civilizations*, a citation from Karl Marx that referred to progress built on "robbing the soil.")

This is not about being anti-innovation or anti-technology or, even, anti-all-speed, per se. This is about being anti- the kind of befuddlement that—as David Orr so eloquently evoked—arises in the wake of information and markets and wealth cut loose from nature. This is about recognizing the need for and working diligently to preserve and restore the possibilities of slow, and the possibilities that we may thereby reintroduce into culture. If there is a need to publish a defense of innovation, then surely there is a need to publish a defense not just of food and farms, but of the slow and of the soil.

In this defense, we must be pro-local and pro-diversity. We must be pro-small. We must on occasion be willing to go beyond the probiotic (one of the latest buzz words in the science of nutritionism and the industry of nutraceuticals) all the way to the pro-poetic.

And, in the end, we must be pro-earthworm.

I cannot end, however, without referring back to what may have struck some readers as an almost snide, somewhat elitist remark made many pages ago: "There is nothing beautiful about the parking lot of a McDonald's."

There is much that is beautiful in using the power of markets to make food more affordable, although there would be something even more beautiful in making more people able to afford good food—fresh food, healthy food, fair food. There is also much, at the other end of the economic continuum, that is beautiful in the dining rooms of Chez Panisse and Blue Hill. But this beauty is incomplete, as incomplete as capitalism itself. For how can we enjoy our slow food while our money is zooming around the planet, investing in companies of the monoculture? If Plato were around, I'm guessing he'd find the greatest beauty neither in the exquisite, expensive ounce of local pheasant nor the tasty, greasy, and affordable Big Mac.

To mediate between these extremes, to find a more beautiful path, and, in the end, a more sustainable and satisfying path, we will have to find the gumption to reach up into cyberspace, clutch as many of our dollars as we can (who knows how much time we have before that Great Game Show Host in the Sky rings his buzzer?), and pull them back down to earth, where we can more truly and honestly and completely put them to work.

acknowledgments

The territories where venture capital, philanthropy, and social investing meet, and, beyond them, the fields of Slow Money, are populated by many wonderful folks without whose help this journey would not have been possible.

Thank you, Jim Lynch, for becoming a freelance photographer after HBS and for giving me my introduction to the world of the Green Revolution, only to find yourself a venture capitalist after all.

Thank you, Jim Fordyce, for teaching me venture capital. My bad investments are mine, all mine.

Thank you, Don Collat, for the acuity of your financial insights and your willingness to venture toward the green.

Thank you, Stuart Davidson, for that day you came up to me a decade ago and said you wanted to help.

Thank you, Hazel Henderson, for your dazzling brand of intellectual heroism and planetary awareness.

Thank you, Noyes family and Steve Viederman, for allowing me to participate in the journey toward dissonance reduction.

Thank you, Ralph Taylor, for a great and wonderful belly laugh and for the commitments you and your family have made to support this work.

Thank you, Matt Sanford, for urging me to speak from the heart that day in Duluth.

Thank you, Lee and Tharon Dunn, for the kind of friendship that makes the word friendship seem woefully inadequate.

Thank you, Gregory Whitehead, for peering into the economic depths in such a profoundly humanistic way.

Thank you, Carlo Petrini. Thank you. . . . No, I need to say more. In your vision, grace, dignity, boldness, and warmth lies something as important as the most heirloom of heirloom varieties, something as profoundly hopeful as an earthworm or a snail.

Thank you, John Fullerton, for allowing your search for the Purpose of Capital to grace the discussions and programs of Investors' Circle.

Thank you, Investors' Circle members, for building such a vital, creative proving ground, marketplace, and community, both for for-profit social entrepreneurs and the investors who wish to support them. There would be no Slow Money without you.

To Slow Money Alliance members go the gratitude of a nurture capitalist and deep appreciation for your willingness to participate in this sometimes-too-audacious-seeming-but-in-the-end-we-trust-just-audacious-enough effort to put the seed back into seed capital. Thanks to you all: Grant Abert, Larry Bain, Jim Baird, Ian and Margo Baldwin, Peter Barnes, Eric Becker, Bill Benenson, Cathy L. Berry, Mary Burns, Claudia P. Casey, Jim Cochran, Eliot Coleman, Tim Crosby, Paul Cultrera, Clifford C. David, Jr., Amy Dickie, Michael Dimock, Paul Dolan, Cathleen Dorinson, Penelope Douglas, Sid Dubose, Bob Estrin, David Feinberg, Theodosia Ferguson, Rian Fried, Wade Greene, Joan Gussow, Larry Jacobs, Michael Kanter, Patty Kestin, Peter D. Kinder, Frederick L. Kirschenmann, Jylle Lardaro, Christopher Lindstrom, Kristin Martinez, Carol Master, Ken Merritt, Diane Edgerton Miller, Tom Miller, Carol Newell and Joel Solomon, Martin Ping,

Matt Reynolds, Simon Rich, Peter Rogers, Richard Rominger, Jeff Rosen, Matt Sanford, Don Shaffer, George Siemon, Lee Slaff, Janice St. Onge, Tom Stearns, Greg Steltenpohl, Tim Storrow, Frank van Beuningen and Margaret McGovern, Gregory Whitehead, Judy Wicks, Tom Willits, and Christiana Wyly.

Thank you, Anne Lennartz.

Thank you, Ruth Reynolds, for pretending to be my hillbilly grandma.

Thank you, Tom Miller, for being my actual hillbilly friend. I wish you could write better.

Thank you, Cathy Berry, for support on so many levels, without which none of this would have been possible.

To Ian and Margo Baldwin, my regrets that those of us of octagonal persuasions and off-the-grid tendencies cannot manage to break bread more often.

Thank you, Jack Lazor, for putting so much inspiration and terroir in a yogurt container.

Thank you, Eliot Coleman, for being so just plain over-the-top fantastically, poetically, intelligently, and beautifully entrepreneurial. And a damned good soil builder.

Thank you, Joan Gussow, for your rare intelligence and steady guidance.

Thank you, Mary Burns, for giving new meaning to the term angel investor.

Thank you, Christine Silverstein, for the kind of activism no community should be without.

Thank you, Z, for the hat trick, and so much, so much more.

Thank you, Zander and Zoe, for putting up with my distraction and confusion.

I could say, "Thank you, Mr. Bartner, for bringing, in Ben

Cohen's words, your heart and mind with you to work everyday," but that would be woefully insufficient.

I could say, "Thank you, Dominic Kulik, for your unparalleled brand of meta-irrational exuberance," but that would fail to take into account all of the times that I have been right and you have been wrong.

I could say, "Thank you, Hubbell, for a seemingly inexhaustible supply of good will, optimism, curiosity and compassion." I think I will.

SLOW MONEY PRINCIPLES

In order to enhance food security, food safety, and food access; improve nutrition and health; promote cultural, ecological, and economic diversity; and accelerate the transition from an economy based on extraction and consumption to an economy based on preservation and restoration, we do hereby affirm the following Principles:

I. We must bring money back down to earth.

II. There is such a thing as money that is too fast, companies that are too big, finance that is too complex. Therefore, we must slow our money down—not all of it, of course, but enough to matter.

III. The twentieth century was the era of Buy Low/Sell High and Wealth Now/Philanthropy Later—what one venture capitalist called "the largest legal accumulation of wealth in history." The twenty-first century will be the era of *nurture capital,* built around principles of carrying capacity, care of the commons, sense of place and nonviolence.

IV. We must learn to invest as if food, farms, and fertility mattered. We must connect investors to the places where they live, creating vital relationships and new sources of capital for small food enterprises.

V. Let us celebrate the new generation of entrepreneurs, consumers, and investors who are showing the way from Making A Killing to Making a Living.

VI. Paul Newman said, "I just happen to think that in life we need to be a little like the farmer who puts back into the soil what he takes out." Recognizing the wisdom of these words, let us begin rebuilding our economy from the ground up, asking:

- *What would the world be like if we invested 50% of our assets within 50 miles of where we live?*
- *What if there were a new generation of companies that gave away 50% of their profits?*
- *What if there were 50% more organic matter in our soil 50 years from now?*

To sign the Slow Money Principles, go to:

www.slowmoneyalliance.org